Biochemistry of Milk Products

Biochemistry of Milk Products

Edited by

A. T. Andrews
Cardiff Institute of Higher Education

J. Varley
University of Reading

The Proceedings of a Symposium organised by the Food Chemistry Group,
RSC Industrial Division, held on 14 December 1993 at Reading University

Special Publication No. 150

ISBN 0-85186-702-2

A catalogue record for this book is available from the British Library

© The Royal Society of Chemistry 1994

All Rights Reserved
*No part of this book may be reproduced or transmitted in any form or
by any means—graphic, electronic, including photocopying, recording,
taping, or information storage and retrieval systems—without written
permission from The Royal Society of Chemistry*

Published by The Royal Society of Chemistry,
Thomas Graham House, Science Park, Cambridge
CB4 4WF

Printed and bound by Bookcraft (Bath) Ltd.

Preface

The milk and dairy products sector forms a very major part of the whole food and agriculture industries, which worldwide take up the greatest share of all human activity. A large proportion of food and drink is consumed directly with little or no pretreatment and a great deal more following only minimal pretreatment, such as peeling, slicing, chopping, heating, etc. These are usually regarded as simple preparation steps and not thought of as manufacturing processes, so only quite a small percentage would generally be thought of as "processed". In this the dairying area is no exception and most milk is still consumed as such rather than being made into products. Nevertheless processing even only a part of the total milk production still represents an extremely important activity. It is worth remembering that historically the oldest manufacturing industries of all were geared to food and drink products, namely fermentation to give alcoholic beverages and the production of cheese from milk. Both of these industries function world wide and are performed on scales varying from amateur operations in the home, through small often specialist cottage industries, to major multinational companies with turnovers of hundreds of millions of dollars.

Cheesemaking began 8-9000 years ago, probably in the Middle East where the stomachs of animals were kept after slaughter and used as leather-type bottles for storing and transporting liquids. It is thought that the storage of milk in imperfectly cleaned stomachs which still contained traces of the digestive enzyme pepsin, aided perhaps by contaminating lactic bacteria, led to clotting of the milk and to the realisation that the resulting curds represented a convenient and concentrated form of most of the protein. Also because of the preservative action of the low pH generated by the fermentation of lactose to lactic acid the curds could be stored for considerable periods of time. In spite of this long history it is still cheesemaking that is the most active and fruitful area in dairy research today. This is very apparent from the contents of this book which represents the proceedings of a recent symposium on advances in dairy biochemistry. There are two principal topics in cheese research at present. Firstly, improvements in starter microorganisms to give better quality cheeses with superior flavour and texture, preferably developed in a shorter time to minimise expensive storage, and to facilitate the tailoring of cheese flavour to particular products and processes. Secondly, the production of new milk coagulants as alternatives to traditional rennet. Both of these topics nowadays lean heavily on molecular biological techniques, the former to develop new micoorganisms with enzyme profiles (peptidases, proteinases and to a lesser degree lipases) better suited to their tasks than current natural microorganisms, and the latter to produce from microorganisms, following genetic manipulation and expression, a purified proteinase of high specificity capable of coagulating milk without the formation of undesirable by-products such as bitter peptides.

In order to be successful in these objectives it is necessary to understand in fine detail at the molecular level all aspects of the cheesemaking process and of what takes place during a typical maturation, including especially the role of starter enzymes, residual coagulant enzymes and indigenous milk enzymes. The initial papers in this symposium were devoted to reviewing and extending knowledge in this area.

The second biggest field in current dairy research covers the functional behaviour of milk proteins. These have been used for many years as ingredients in a wide variety food products and in dietetic applications because of their desirable physical attributes and high nutritional quality. Only recently however have separation methods improved to the extent that the large-scale production of individual protein components with specific functional properties for particular applications has become a commercially viable route to new food ingredients. Again, such advances depend upon a thorough understanding at the molecular level of the processes involved, in this case of what molecular features make a protein a good emulsifying agent, enable it to form gels, or to stabilise foams, etc. Once these features can be identified it then becomes a practical proposition to alter proteins via the genetic engineering route to enhance or modify functional behaviour and so ultimately to produce tailor-made proteins designed to fulfil a particular task. For these reasons other invited papers in this symposium and a number of the supporting poster presentations cover not only the production and functional evaluation of natural proteins but also the effect that genetic substitution of particular amino acid residues has on functional behaviour. This should give us a greater understanding of the interactions involved, which should in turn lead later to proteins with improved performance.

Inevitably, as the proceedings of what was only a one-day symposium the selection of topics covered may appear to be somewhat limited, but we believe that the most active areas of dairy research are well represented and that the review-like nature of most of the invited papers means that the coverage is much less limited than would be expected. Many of the findings described, and certainly the techniques used, will undoubtedly be applicable not only elsewhere in the milk and dairy chemistry area but also widely outside it, especially in the microbiological and protein chemistry/protein engineering fields. These proceedings give a good state-of-the-art picture of current research which should be very valuable to research workers, lecturers, graduate students and final year undergraduates with interests in the practical applications of molecular biology, enzymology and protein chemistry, not just in improving the quality and performance of dairy foods and ingredients but also in a much wider context.

We should like to thank all those who made this symposium possible by both their physical and unflagging moral support, and especially all the contributors of papers and posters whose universally excellent quality manuscripts made our task as editors so straightforward and enjoyable.

A.T.Andrews
J.Varley

Contents

Proteolysis in Cheese during Ripening   1
   P.F.Fox, T.K.Singh and P.L.H.McSweeney

Manipulation of Proteolysis in *Lactococcus Lactis*   32
   A.J.Haandrikman, I.Mierau, J.Law, K.J.Leenhouts, J.Kok and
   G.Venema

New Starter Cultures for Cheese Ripening   47
   B.A.Law

Engineering Pivotal Proteins for Lactococcal Proteolysis   56
   W.M.de Vos and R.J.Siezen

Protein Engineering and Preliminary X-ray Analysis of CHY155-165RHI   72
Loop Exchange Mutant
   J.E.Pitts, P.Orprayoon, P.Nugent, R.V.Dhanaraj, J.B.Cooper,
   T.L.Blundell, J.Uusitalo and M.Penttila

Peptidases from Lactococci and Secondary Proteolysis of Milk Proteins   83
   F.Mulholland

Functional Milk Protein Products   94
   D.M.Mulvihill

Protein Engineering Studies of β-Lactoglobulin   114
   L.Sawyer, J.H.Morais Cabral and C.A.Batt

Functional Properties of Chhana Whey Products   127
   A.S.Grandison and A.R.Jindal

Thermal Aggregation of Whey Protein Concentrates under Fluid Shear   133
Conditions
   A.J.Steventon, A.M.Donald and L.F.Gladden

Debittering of α-Casein Hydrolysates by a Fungal Peptidase   143
   J.Gallagher, A.D.Kanekanian and E.P.Evans

The Effect of Thermisation on the Thermal Denaturation of 152
γ-Glutamyltranspeptidase in Milk and Milk Products
S.S.Patel and R.A.Wilbey

Keeping Quality of Pasteurised and High Pasteurised Milk 157
B.Borde-Lekona, M.J.Lewis and W.F.Harrigan

Fouling and UHT Processing 162
P.Kastanas, M.J.Lewis and A.Grandison

Ultrafiltration of Sweet Cream Buttermilk 169
H.G.Ramachandra Rao, M.J.Lewis and A.S.Grandison

Index 177

Proteolysis in Cheese during Ripening

P. F. Fox, T. K. Singh, and P. L. H. McSweeney

DEPARTMENT OF FOOD CHEMISTRY, NATIONAL FOOD BIOTECHNOLOGY CENTRE, UNIVERSITY COLLEGE, CORK, IRELAND

1. INTRODUCTION

The conversion of milk to cheese curd is only the first stage in the production of most cheese varieties. Essentially all hard, and many soft, cheeses are ripened for periods ranging from a few weeks to two years or longer. During this period, cheeses undergo numerous biochemical changes which lead to the development of the appropriate texture, flavour and aroma.

The biochemistry of cheese ripening is very complex, among the most complex of any food. It involves 3 primary processes: glycolysis, lipolysis and proteolysis, the relative importance of which depends on the variety. Proteolysis is the most complex of these phenomena and has been the subject of much research in recent years. Methods for assessing proteolysis have been reviewed extensively[1-5] and will not be discussed further here.

2. CHEESE RIPENING AGENTS AND THEIR CONTRIBUTION TO PROTEOLYSIS

The extent of proteolysis in cheese varies from very limited (e.g. Mozzarella) to very extensive (e.g. blue mould varieties). The products of proteolysis range in size from large polypeptides, comparable in size to intact caseins, through a range of medium and small peptides to free amino acids.

Proteolytic agents in cheese generally originate from 5 sources: the coagulant, the milk, starter bacteria, non-starter bacteria and adjunct starter. Enzymes from the first four sources are active in nearly all ripened cheeses. The adjunct starter (i.e. microorganisms added to cheesemilk for purposes other than acidification) can exert considerable influence on maturation in cheese varieties in which they are used (e.g. *Penicillium roqueforti*, *P. camemberti* in mould-ripened varieties or *Brevibacterium linens* in smear-ripened cheeses). Exogenous enzymes used to accelerate ripening could be added to the above list, and when present can be very influential.

Quantitation of the contribution of these agents individually or in various combinations has been attempted using three complimentary approaches: (1) model cheese systems from which the non-starter microflora have been eliminated by aseptic techniques, in which acidification may be accomplished using an acidogen (usually gluconic acid-δ-lactone) rather than starter, and in which coagulant and indigenous milk enzymes may be inactivated or inhibited; these techniques have been reviewed,[6]

(2) activity and specificity of the principal proteinases and peptidases on caseins or casein-derived peptides in solution, and (3) isolation of peptides from cheese and, based on the known specificity of the proteinases/peptidases on the caseins in solution, identificiation of their formative agent(s) in cheese.

The use of model systems has enabled the contribution of the principal ripening agents to proteolysis in cheese to be estimated fairly precisely. The subject has been reviewed[6] and may be summarized as follows. Using aseptic control cheeses, aseptic rennet-free cheeses, aseptic starter-free cheeses and aseptic, rennet-free, starter-free cheeses, Visser[7-10] and Visser and de Groot-Mostert[11] concluded that the coagulant is responsible for the initial hydrolysis of caseins, e.g. as shown by PAGE and the production of most of the water- or pH 4.6-soluble N in Gouda-type cheese, and that the actions of indigenous milk and starter proteinases are less important at this level of proteolysis. However, the production of small peptides and amino acids is due primarily to the action of starter bacteria or their enzymes. The results of other studies[12-17] on controlled-microflora cheese yielded generally similar results.

Direct evidence for the rôle of the principal indigenous milk proteinase, plasmin, in cheese is limited. Visser[10] found that approximately 5% of the total N in a 6 month-old aseptic starter-free, rennet-free cheese was soluble at pH 4.6, but very low levels of free amino acids were found. Farkye and Fox,[18] who eliminated the influence of plasmin in Cheddar cheese by an inhibitor, 6-aminohexanoic acid (AHA), found differences between electrophoretograms of these cheeses and those of controls; bands corresponding to the γ-caseins (produced from β-casein by plasmin) were less intense in cheese containing AHA. These authors also found that AHA reduced the level of water-soluble N, again suggesting a rôle for plasmin in the initial hydrolysis of caseins.

Milk also contains an indigenous acid proteinase (pH optimum, 4.0) which is probably cathepsin D (E.C. 3.4.23.5).[19-22] Cathepsin D is relatively heat-labile and is probably largely inactivated by pasteurization. The action of cathepsin D on the caseins is very similar to that of chymosin,[21, 23] and it has been suggested that a band with an electrophoretic mobility corresponding to α_{s1}-CN f24-199, i.e. the primary product of chymosin action on α_{s1}-casein, in aseptic Meshanger-type cheese is due to its action.[24] The contribution of cathepsin D to proteolysis in cheese has not been quantified.

Although non-starter lactic acid bacteria (NSLAB) usually dominate the microflora of Cheddar-type cheese during much of its ripening (see ref. 25), their influence on proteolysis in cheese has been neglected by most authors. Visser[8-10] used an aseptic control cheese to eliminate the influence of the NSLAB, as did Desmazeaud et al.[14] and O'Keeffe et al.[15-17] A wide range of proteolytic enzymes has been identified in NSLAB (see ref. 25), and thus it is likely that they play a rôle in proteolysis in cheese. In a comparative study on the ripening characteristics of Cheddar cheese made from raw, pasteurized or microfiltered milks, McSweeney et al.[26] concluded that non-starter lactobacilli were responsible for differences in proteolysis in the cheese made from the raw milk, particularly with regard to the formation of short peptides and free amino acids.

The progress of proteolysis in most ripened cheeses can be summarized as follows:-   initial hydrolysis of caseins is caused primarily by residual coagulant, and to a lesser extent by plasmin and perhaps cathepsin D, resulting in the formation of large and intermediate-sized peptides which are subsequently degraded by the coagulant and enzymes from the starter and non-starter flora. The production of small peptides and free amino acids results from the action of bacterial proteinases and peptidases. This general outline of proteolysis can vary substantially between varieties where differences in manufacturing practices can have a profound influence on proteolysis. In Mozzarella, Swiss and other high-cook varieties, coagulant is extensively or completely denatured by the high cooking temperature. In these varieties, the contribution of plasmin to the initial hydrolysis of caseins is more pronounced than in Cheddar and Dutch varieties. Likewise, in mould or bacterial surface-ripened varieties, proteinases and peptidases from the adjunct starter influence proteolysis strongly.

This presentation will focus on the specificity of the principal proteinases and peptidases in cheese on the individual caseins and casein-derived peptides and on the isolation and identification of peptides from Cheddar cheese.

3. SPECIFICITY OF THE PRINCIPAL PROTEINASES AND PEPTIDASES IN CHEESE

3.1. Proteinases from the Coagulant

Chymosin (E.C. 3.4.23.4) is the principal proteinase in traditional rennets used for cheesemaking. It is an aspartyl proteinase of gastric origin, secreted by the young of a number of mammalian species. The principal rôle of chymosin in cheesemaking is to coagulate the milk. However, about 6% of the chymosin added to cheese milk is retained in the curd for Cheddar and plays a major rôle in the initial proteolysis of caseins in many cheese varieties (see Section 2).

The action of chymosin on the B-chain of insulin indicates that it is specific for hydrophobic and aromatic amino acid residues.[27] Relative to many other proteinases, chymosin is weakly proteolytic; indeed, limited proteolysis is one of the characteristics to be considered in the selection of proteinases for use as rennet substitutes.[2]

The primary chymosin cleavage site in the milk protein system is the Phe_{105}-Met_{106} bond in κ-casein which is many times more susceptible to chymosin than any other bond in milk proteins; its hydrolysis leads to coagulation of the milk (see ref. 28). Cleavage of κ-casein Phe_{105}-Met_{106} yields para-κ-casein (κ-CN f1-105) and glycomacropeptides (GMPs; κ-CN f106-169). Most of the GMPs are lost in the whey but para-κ-casein remains attached to the casein micelles and is incorporated into the cheese.

Although considerably less susceptible than the Phe_{105}-Met_{106} bond of κ-casein, α_{s1}-, α_{s2}- and β-caseins are readily hydrolysed by chymosin under appropriate conditions. A number of authors have investigated the action of chymosin on β-casein.[29-32] In solution in 0.05 M Na acetate buffer, pH 5.4, chymosin cleaves β-casein at 7 sites: Leu_{192}-Tyr_{193} > Ala_{189}-Phe_{190} > Leu_{165}-Ser_{166} ≥ Gln_{167}-Ser_{168} ≥ Leu_{163}-Ser_{164} > Leu_{139}-Leu_{140} ≥ Leu_{127}-Thr_{128}.[31] The Michaelis parameters, K_m and k_{cat}, for the action of chymosin on the bond Leu_{192}-Tyr_{193} are 0.075 mM and 1.54 s^{-1}, respectively, for micellar β-casein and 0.007 mM and 0.56 s^{-1} for the monomeric protein.[32] NaCl inhibits the hydrolysis of β-casein by chymosin to an extent dependent on concentration and pH; the relative rates of hydrolysis of certain peptide bonds by chymosin are influenced by the ionic conditions of the solution.[33]

The primary site of chymosin action on α_{s1}-casein is Phe_{23}-Phe_{24}.[34, 35] Cleavage at this site has significance in producing a small peptide (α_{s1}-CN f1-23) which is further hydrolyzed by starter proteinases, and in the softening of cheese texture.[36] The specificity of chymosin on α_{s1}-casein in solution has been studied.[29,37-39] In 0.1 M phosphate buffer, pH 6.5, chymosin cleaves α_{s1}-casein at Phe_{23}-Phe_{24}, Phe_{28}-Pro_{29}, Leu_{40}-Ser_{41}(?), Leu_{149}-Phe_{150}, Phe_{153}-Tyr_{154}, Leu_{156}-Asp_{157}, Tyr_{159}-Pro_{160} and Trp_{164}-Tyr_{165}.[39] These sites are also cleaved at pH 5.2 in the presence of 5% NaCl and, in addition, Leu_{11}-Pro_{12}, Phe_{32}-Gly_{33}, Leu_{101}-Lys_{102}, Leu_{142}-Ala_{143} and Phe_{179}-Ser_{180}. The rates at which specific peptide bonds appear to be cleaved are dependent on the ionic conditions and differ between pH 6.5 and pH 5.2 in the presence of 5% NaCl, particularly Leu_{101}-Lys_{102}, which is cleaved far faster at the lower pH.[39]

Carles and Ribadeau-Dumas[35] found that the k_{cat} and K_m for chymosin at pH 6.2 and 30°C on the Phe$_{23}$-Phe$_{24}$ bond of α_{s1}-casein were 0.66 s^{-1} and 0.37 mM, respectively. The influence of pH and urea on the hydrolysis of α_{s1}-casein by chymosin was studied by Mulvihill and Fox[40] who found that pH affected the pattern of proteolysis. Ionic conditions also affected proteolysis.[37] Dunn et al. [41] reported that the hydrolysis of synthetic octapeptides of the type Lys-Pro-Xxx-Yyy-Phe-(NO$_2$)Phe-Arg-Leu by chymosin is pH-dependent.

α_{s2}-Casein appears to be relatively resistant to proteolysis by chymosin but the specificity of chymosin, and indeed of other proteinases, on this protein has received little attention. Chymosin cleavage sites in α_{s2}-casein (pH 6.5) are generally restricted to the hydrophobic regions of the molecule, i.e. residues 90-120 and 160-207: Phe$_{88}$-Tyr$_{89}$, Tyr$_{95}$-Leu$_{96}$, Gln$_{97}$-Tyr$_{98}$, Tyr$_{98}$-Leu$_{99}$, Leu$_{99}$-Tyr$_{100}$, Phe$_{163}$-Leu$_{164}$, Phe$_{174}$-Ala$_{175}$ and Tyr$_{179}$-Leu$_{180}$; the primary site appears to be Phe$_{88}$-Tyr$_{89}$.[42]

The extent of chymosin action on para-κ-casein (κ-CN f1-105) is unclear. Para-κ-casein migrates in the opposite direction to the other caseins in the alkaline urea PAGE gels widely used to study the initial proteolysis of caseins during ripening and thus is often ignored in such studies. Although it contains a number of potential chymosin cleavage sites, Green and Foster[43] found that para-κ-casein is resistant to chymosin action.

The parameter, k_{cat}/K_m, a measure of the affinity of an enzyme for a substrate, for the action of chymosin on the most susceptible bonds in κ-, β- and α_{s1}-caseins (Phe$_{105}$-Met$_{106}$, Leu$_{192}$-Tyr$_{193}$ and Phe$_{23}$-Phe$_{24}$, respectively) has been estimated as 1405, 20.6 and 1.8 s^{-1} mM^{-1}, respectively,[32, 35] which suggests that the second most susceptible bond in the caseins to chymosin action is Leu$_{192}$-Tyr$_{193}$ of β-casein.

Calf rennet contains about 10% bovine pepsin (E.C. 3.4.23.1) which contributes to proteolysis in Cheddar cheese;[44, 45] cleavage of the bond Leu$_{109}$-Glu$_{110}$ in α_{s1}-casein appears to be due to its action.[45] Peptides produced from Na caseinate[46] or β-casein[47] by bovine pepsin appear to be generally similar to those produced by chymosin, although, as far as we are aware, the specificity of bovine or porcine pepsins on bovine caseins has not been determined rigorously.

For several years, the supply of calf rennet has been insufficient to meet demand and much effort has been expended on searching for suitable rennet substitutes (see refs. 48 and 49). Several enzymes have been studied, including bovine,[50] porcine,[46] ovine,[51] and chicken[52] pepsins. Microbial proteinases studied include those from *Cryphonectria parasitica*, *Rhizomucor pusillus*, *R. miehei*, *Penicillium janthinellum*, *Rhizopus chinensis* and *Aspergillus usameii*; the first 3 are used commercially, especially in the USA. The specificity of many of these enzymes on the oxidized B-chain of insulin was summarized by Green.[48] Their specificities on α_{s1}- and β-caseins differ from that of chymosin[49] but have not been established. The introduction of recombinant chymosins has limited the use of these enzymes.

Recombinant calf chymosins, expressed by *Aspergillus niger* var. *awamori*, *Kluveromyces marxianus* var. *lactis* and *Escherischia coli*, were introduced recently and, since their acceptance by regulatory authorities for use in cheese, they have been used widely for cheesemaking in many, but not all, countries. Cheesemaking trials, involving a number of varieties, have shown only small differences between cheese made with calf rennet or recombinant chymosins.[44, 53-57] Recombinant chymosins may contain only one genetic variant,[58] while calf rennet can contain three chymosin variants, A, B and C,[59] as well as bovine pepsin. Possible differences in specificity between chymosin variants have not been reported.

Residual coagulant activity in cheese is a function of a number of technological parameters, including the pH of the curd at whey drainage (which influences the amount of coagulant retained in the curd), final pH of the cheese and, especially, the cook temperature. Cheeses which are cooked at a high temperature (e.g. Mozzarella and Swiss varieties) have relatively little coagulant activity. Chymosin and other enzymes from the coagulant act primarily on α_{s1}-casein in cheese. β-Casein is a good substrate for chymosin in solution but not in cheese, perhaps because the hydrophobic C-terminal region of the protein, which contains chymosin-susceptible sites, may be involved in hydrophobic interactions. Chymosin is probably inactive, or very weakly active, on α_{s2}-casein in cheese as this protein is not a good substrate for chymosin in solution.

3.2. Indigenous Milk Proteinases

The presence of indigenous proteinases in milk has been recognized for nearly a century. The principal indigenous proteinase is plasmin, which is optimally active at pH 7.5, while the lesser, cathepsin D, has a pH optimum at ~4.0.

3.2.1. Plasmin. Plasmin (fibrinolysin, E.C. 3.4.21.7) has been the subject of much study (for review see ref. 60). The physiological rôle of plasmin is solubilization of fibrin clots, to achieve which there exists a complex fibrinolytic system consisting of the active enzyme, its zymogen, activators and inhibitors of the enzyme and zymogen activators, all of which are present in milk. Plasmin, plasminogen and plasminogen activators are associated with the casein micelles in milk, while the inhibitors are in the serum phase.[60, 61]

Plasmin, a trypsin-like proteinase with a high specificity for peptide bonds containing lysyl residues, is active on all caseins, but especially α_{s2}- and β-caseins.[60]

Plasmin cleaves β-casein in solution at 5 principal sites (Figure 1): Lys_{28}-Lys_{29}, Lys_{105}-His_{106}, Lys_{107}-Glu_{108}, Lys_{113}-Tyr_{114} and Arg_{183}-Asp_{184} with the formation of the polypeptides, γ_1-CN (β-CN f29-209), γ_2-CN (f106-209), γ_3-CN (f108-209), γ_4-CN (f114-209) and γ_5-CN (β-CN f184-209) and protease peptone 5 (β-CN f1-105 and f1-107), protease peptone 8-slow (β-CN f29-105 and f29-107), protease peptone 8-fast

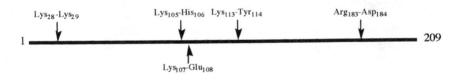

PRODUCTS

Proteose Peptones		γ-Caseins
Known	*Probable*	
PP8f β-CN f1-28	β-CN f106-113	γ^1-CN (β-CN f29-209)
PP8s β-CN f29-105	β-CN f108-113	γ^2-CN (β-CN f106-209)
PP8s β-CN f29-107	β-CN f114-183	γ^3-CN (β-CN f108-209)
PP-T β-CN f29-113	β-CN f106-183	γ^4-CN (β-CN f114-209)
PP5 β-CN f1-105	β-CN f108-183	γ^5-CN (β-CN f184-209)
PP5 β-CN f1-107	β-CN f1-113 (unlikely)	

Figure 1   Specificity of plasmin on β-casein[62, 66, 67] (PP8f, protease peptone 8 fast; PP8s, protease peptone 8 slow; PP-T, protease peptone T; PP5, protease peptone 5).

(β-CN f1-28) and proteose peptone T (β-CN f29-113?) and probably the fragments β-CN f113-183, f106-113 and f108-113.[62-67]

Plasmin hydrolyses α_{s2}-casein in solution at 8 sites: Lys_{21}-Gln_{22}, Lys_{24}-Asn_{25}, Arg_{114}-Asn_{115}, Lys_{149}-Lys_{150}, Lys_{150}-Thr_{151}, Lys_{181}-Thr_{182}, Lys_{188}-Ala_{189} and Lys_{197}-Thr_{198},[68, 69] producing about 14 peptides, three of which are potentially bitter.[69]

Although plasmin is less active on α_{s1}- than on α_{s2}- or β-casein, the formation of λ-casein, a minor casein component, has been attributed to its action on α_{s1}-casein.[70] The principal plasmin cleavage sites in α_{s1}-casein are: Arg_{22}-Phe_{23}, Arg_{90}-Tyr_{91}, Lys_{102}-Lys_{103}, Lys_{103}-Tyr_{104}, Lys_{105}-Val_{106}, Lys_{124}-Glu_{125} and Arg_{151}-Gln_{152}.[71]

Although there are a number of potential plasmin cleavage sites in κ-casein, plasmin has low activity on this protein and its specificity does not appear to have been determined. Eigel[72] found no hydrolysis of κ-casein under conditions adequate for the complete hydrolysis of α_{s1}-casein, but Andrews and Alichanidis[65] reported that hydrolysis of κ-casein by plasmin accounted for 4% of the proteose peptone fraction produced by indigenous plasmin in pasteurized milk stored at 37°C for 7 days and detectable by PAGE.

3.2.2. Other Indigenous Milk Proteinses.

The specificity of cathepsin D on the caseins has not been determined but electrophoretograms of caseins incubated with milk acid proteinase or exogenous cathepsin D indicate a specificity very similar to that of chymosin[19, 23] although the enzymes differed with respect to rates of cleavage of certain peptide bonds. Interestingly, κ-casein appears to be a poor substrate for cathepsin D which has very poor milk clotting properties.

The presence of other minor proteolytic enzymes in milk has been reported, including thrombin,[73] a lysine aminopeptidase[73] and proteinases from leucocytes,[74, 75] but they are considered not to be very significant.[74, 60] The occurrence of cathepsin D in milk suggests that other lysosomal proteinases are also present, although, as far as we are aware, none of these has yet been detected in milk.

3.3. Combined Action of Chymosin and Plasmin

The theoretical combined action of chymosin and plasmin on the principal cleavage sites of α_{s1}- and β-caseins is shown schematically in Figure 2. Theoretically,

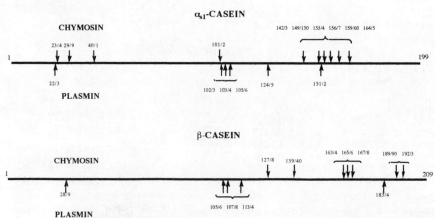

Figure 2   Potential combined action of chymosin and plasmin in cheese

their combined action could produce quite small peptides and their specificities are in fact complementary, especially on β-casein which chymosin cleaves primarily toward the C-terminal while plasmin cleaves mainly in the N-terminal region. As far as we know, the combined action of these enzymes on the isolated caseins has not been studied and it is not known whether they act in a concerted manner in cheese.

3.4. Proteolytic Enzymes from Starter

3.4.1. Proteinases of Lactococcus and Lactobacillus The starter cells are a major source of proteinases and peptidases in cheese. The proteolytic system of *Lactococcus*, the most widely used cheese starter, has been studied thoroughly, while those of thermophilic *Streptococcus* and *Lactobacillus* starters have recently attracted considerable attention.

The principal proteinase of the lactic acid bacteria is associated with the cell wall, where it has ready access to extracellular proteins. Cell wall-associated proteinases of *Lactococcus* can be classified into 3 groups, P_I-, P_{III}- and P_I/P_{III}-types. P_I-type proteinases degrade β- but not $α_{s1}$-casein at a significant rate, P_{III}-type proteinases readily hydrolyse both $α_{s1}$- and β-caseins,[76] while P_I/P_{III}-types have intermediate specificity. The nucleotide sequences of the genes for both P_I- and P_{III}-type proteinases have been established;[77-79] few differences are apparent and alteration of a few residues by site-directed mutagenesis can alter the specificity of the proteinase.[80] Exterkate[81] reported that a cation-binding site in P_I proteinases, but absent in P_{III}, was mainly responsible for the difference in specificity between the enzymes on positively-charged chromophoric peptides. The specificities of cell wall-associated proteinases from a number of strains of *Lactococcus* on $α_{s1}$-, $α_{s2}$-, β- and κ-caseins have been determined (Figures 3 to 5; refs. 82-88). Tan *et al.*[89] commented that P_I-type proteinases have a rather broad specificity on β-casein and that its cleavage

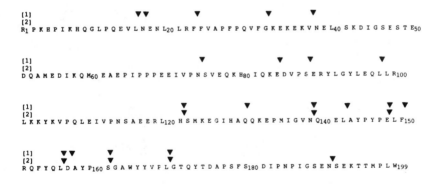

Figure 3   Amino acid sequence of *Bos* $α_{s1}$-casein showing the position of the cleavage sites of cell wall-associated proteinases of [1] *Lactococcus lactis* ssp. *cremoris* SK112 (ref. 87) and [2] *L. lactis* ssp. *lactis* NCDO 763 (ref. 84).

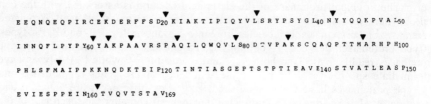

Figure 4   Amino acid sequences of *Bos* α_{s2}-casein A and κ-casein B showing the positions of the cleavage sites of cell wall-associated proteinase of *Lactococcus lactis* ssp. *lactis* NCDO 763 (ref. 84).

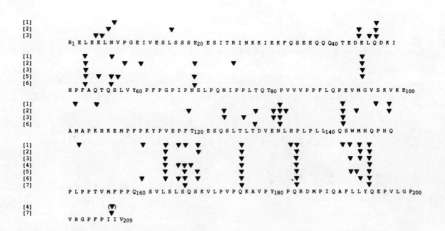

Figure 5   Amino acid sequence of *Bos* β-casein showing the cleavage sites of cell wall-associated proteinases of *Lactococcus*. [1] P_I-type, *L. lactis* ssp. *cremoris* H2 (ref. 88); [2] P_{III}-type, *L. lactis* spp. *cremoris* SK112 (ref. 88); [3] P_{III}-type, *L. lactis* ssp. *cremoris* AM1 (ref. 86); [4] *L. lactis* ssp. *cremoris* HP (ref. 85); [5] *L. lactis* ssp. *cremoris* AC1 (ref. 83); [6] *L. lactis* ssp. *lactis* NCDO 763 (ref. 83); [7] *L. lactis* ssp. *lactis* NCDO 763 (ref. 82).

sites typically contained a Gln or Ser residue and are likely to lie in regions of the molecule which have high hydrophobicity, a high proline content and a low charge. On the other hand, P_{III}-type proteinases tend to cleave bonds of the type Glx-X or X-Glx, where X is generally a large, hydrophobic residue (Met, Phe, Leu or Tyr), while a hydrophobic residue is usually found at the P_2 or P'_2 position with a negatively-charged residue in the P_2-P_3 or P'_2-P'_3 regions. In general, P_{III}-type proteinases have broader specificities on β-casein than P_I-type enzymes,[89] which Visser et al.[86] suggested might explain why such strains produce less bitter peptides from casein than P_I-type strains.[90] The literature on lactococcal cell wall-associated proteinases has been reviewed extensively.[6,89,91] Thermophilic Lactobacillus spp. used as starters also possess a cell wall-associated proteinase (see ref. 91)

The primary rôle of lactococcal proteinases in cheese appears to be the hydrolysis of large and intermediate-sized peptides produced from caseins by chymosin or plasmin. A number of authors have investigated the action of lactococcal cell wall proteinases on such peptides (Figure 6). The cell wall-associated proteinase of Lactococcus does not appear to be important in the initial hydrolysis of β-casein in Cheddar as detected by urea-PAGE, perhaps because the primary cleavage sites on this

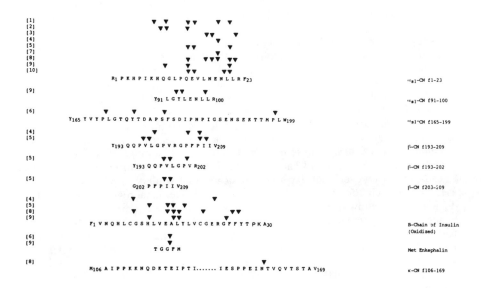

Figure 6   Amino acid sequence of a number of peptides from casein hydrolysates and other sources showing the specificity of lactococcal proteinases and peptidases. [1] Cell wall associated proteinase of L. lactis ssp. cremoris H61 (ref. 96); [2] cell wall-associated proteinase (P_I-type) of L. lactis ssp. cremoris HP (ref. 97); [3] cell wall-associated proteinase (P_{III}-type) of L. lactis ssp. cremoris AM$_1$ (ref. 97); [4] metalloendopeptidase from L. lactis ssp. cremoris HP (ref. 98); [5] neutral oligo-endopeptidase from L. lactis ssp. cremoris C13 (ref. 98); [6] LEP III-I from L. lactis ssp. lactis MG 1363 (ref. 99); [7] LEP I from L. lactis ssp. cremoris (ref. 100); [8] cell wall-associated proteinase of L. lactis ssp. lactis NCDO 763 (ref. 84); [9] LEP II from L. lactis ssp. cremoris H61 (ref. 101); [10] LEP III-I from L. lactis ssp. lactis MG 1363 (ref. 99).

substrate are rendered inaccessible due to hydrophobic interactions of the C-terminal region of the molecule.[92]

3.4.1. Peptidases of *Lactococcus* and *Lactobacillus*

Many of the peptides discussed above are too large to permit their transport into the bacterial cell[93] and further proteolysis is necessary before the cell can utilize the amino acids they contain. *Lactococcus* spp. possess a very wide range of peptidases, *viz.* endo-, amino-, di- and tri-peptidases and proline-specific peptidases, but apparently not a carboxypeptidase. Lactococcal peptidases which have been isolated and characterized are summarized in Table 1; reviews of the relevant literature include refs 89, 91, 94 and 95.

Lactococcal aminopeptidases have an intracellular location. However, PepX (XPDA), tripeptidase and endopeptidases of *Lactococcus* appear to have a peripheral location.[138] The range of peptides which could be produced from β-casein by the combined action of the proteinases and peptidases of *Lactococcus* was described by Smid *et al.*[139] and an example of the degradation of a hypothetical oligopeptide to free amino acids by the combined action of lactococcal peptidases is shown in Figure 7. The action of PepN on peptides produced from β-casein by trypsin was studied by Tan *et al.*.[140] As expected, trypsin hydrolyzed β-casein extensively but PepN was active only on 10 of the resulting peptides. Trypsin and plasmin have similar specificities; if the results of Tan *et al.*[140] are extrapolated to the plasmin cleavage sites on β-casein, PepN is potentially able to hydrolyse only one N-terminal residue from a plasmin-produced peptide, i.e. Glu_{108} from β-CN f108-209 (γ^3-CN), although PepN may not be able to act on a polypeptide as large as γ^3-CN. Further studies of the action of lactococcal endo- and exopeptidases on casein-derived peptides appear to be warranted. The results of such studies may explain the accumulation of certain peptides in cheese and thus how lactococcal peptidases can direct proteolysis. An example is the degradation of α_{s1}-CN f1-23 (produced rapidly by chymosin) by the starter. α_{s1}-Casein f1-23 has been identified only in Cheddar cheese made under aseptic conditions and acidified using gluconic acid-δ-lactone (O'Sullivan and Fox, unpublished) or in cheese made with proteinase-negative starter,[141] although its complementrary polypeptide, α_{s1}-CN f24-199, is a major peptide in young Cheddar. Lactococcal cell wall-associated proteinases appear to hydrolyse α_{s1}-CN f1-23 rapidly to a range of smaller peptides, including α_{s1}-CN f1-9, f1-13 and f11-? which have been identified in Gouda[96] and Cheddar.[142] However, our experience indicates that while α_{s1}-CN f1-9 and f1-13 accumulate, their complementary peptides from the C-terminal region of α_{s1}-CN f1-23 do not, suggesting that the latter peptides are degraded rapidly by intracellular lactococcal exopeptidases. The N-terminal sequence of α_{s1}-CN f1-9 and f1-13 is R.P.H.K.P and therefore they should be susceptible to the action of the lactococcal PepX. R.P *p*-nitroanalides are cleaved by the PepX of *Streptococcus thermophilus*[123] although Booth *et al.*[111] found that the PepX of *Lc. lactis* ssp. *cremoris* AM2 was not active on R.P.P or R.P.P.G.F. Thus, it would appear that, since α_{s1}-CN f1-9 and f1-13 accumulate to become major components of the water extract of Cheddar, PepX is either inactive in cheese or is unable to hydrolyse these peptides, perhaps because of the positively-charged His and Lys residues close to their N-terminus.

Proteinases and Peptidases from the Secondary Starter and Non-Starter Flora

Proteinases and peptidases from the adjunct starter can play important rôles in proteolysis in cheese varieties where such adjuncts are used. A number of authors (e.g. refs. 143, 144) have used lactobacilli as adjuncts in the manufacture of Cheddar; their effect on proteolysis is presumably the same as non-starter lactobacilli, as discussed above. For references on the proteolytic enzymes of traditional adjuncts, ie.,

Table 1. Peptidases of *Lactococcus* and *Lactobacillus*

Peptidase	Strain	Substrate	MW/kDa	Opt. Activity pH	Opt. Activity °C	Subunits	Class	Reference
Endopeptidases								
Lactococcus								
LEP[1]	*Lc. lactis* ssp. *lactis* CNRZ 267	peptides	49	-	-	-	-	102
LEP I	*Lc. lactis* ssp. *cremoris* H61	peptides	98	7-7.5	40	1	metallo	100
LEP II	*Lc. lactis* ssp. *cremoris* H61	peptides	80	6	37	2	metallo	101
LEP III	*Lc. lactis* ssp. *cremoris* Wg2	peptides	70	6-6.5	30-38	1	neutral	103
LEP (MEP[2])	*Lc. lactis* ssp. *cremoris* HP	peptides	180	8-9	42	-	metallo	98
LEP (NOP[3])	*Lc. lactis* ssp. *cremoris* C13	peptides	70	-	-	-	neutral	98
LEP III-1	*Lc. lactis* ssp. *lactis* MG 1363	peptides	70	-	-	-	-	99
Aminopeptidases								
Lactococcus								
AMP[4] I	*Lc.* bv. *diacetylactis* CNRZ 267	Lys-*p*-NA	85	6.5	35	-	metallo	104
AMP III	*Lc. lactis* ssp. *cremoris* AC1	Lys-*p*-NA	36	7	40	1	metallo	105
PepN	*Lc. lactis* ssp. *cremoris* WG2	Lys-*p*-NA	95	7	40	1	metallo	106
GAP[5] I	*Lc. lactis* ssp. *cremoris* HP	Glu-/Asp-*p*-NA	130	-	50-55	3	metallo	107
GAP II	*Lc. lactis* ssp. *lactis* NCDO 712	Glu-*p*-NA	245	8	65°C	6	metallo	108
PepA (GAP III)	*Lc. lactis* ssp. *cremoris* HP	Glu-/Asp-*p*-NA	520	8	50	-	metallo	98
PepX	*Lc. lactis* ssp. *cremoris* P8-2-47	X-Pro-*p*-NA	180	7	45-50	2	serine	109
PepX	*Lc. lactis* ssp. *lactis* NCDO 763	X-Pro-*p*-NA	190	8.5	40-45	2	serine	110
XAP[6] II	*Lc. lactis* ssp. *cremoris* AM2	X-Pro/X-Ala	117	6-9	-	-	serine	111
XAP III	*Lc. lactis* ssp. *lactis* H1	X-Pro-*p*-NA	150	6-9	-	-	serine	112
PCP[7]	*Lc. lactis* ssp. *cremoris* HP	Pyr-*p*-NA	-	-	-	-	-	113
PCP	*Lc. lactis* ssp. *cremoris* HP	Pyr-*p*-NA	80	8-8.5	37	2	serine	98
PepC	*Lc. lactis* ssp. *cremoris* AM2	His-β-NA	300	7	40	6	thiol	114

1. LEP, lactococcal oligoendopeptidase; 2. MEP, metalloendopeptidase. 3. NOP neutral oligoendopeptidase. 4. AMP, general amino peptidase. 5. GAP, glutamyl aminopeptidase. 6. XAP, X-prolyl-dipeptidyl aminopeptidase. 7. PCP, pyrrolidonyl carboxyl peptidase

Table 1. Cont'd.

Peptidase	Strain	Substrate	MW/kDa	Opt. Activity pH	Opt. Activity °C	Subunits	Class	Reference
Aminopeptidases								
Lactobacillus								
AMP[4] II	*Lb. delbrueckii* ssp. *lactis* 1183	Lys-*p*-NA	78-91	6.2-7.2	47.5	1	metallo	115
AMP III	*Lb. acidophilus* R-26	Lys-*p*-NA	38	-	-	-	metallo	116
AMP IV	*Lb. delbrueckii* ssp. *bulgaricus* CNRZ 397	Lys-*p*-NA	95	-	-	-	metallo	117
AMP V	*Lb. helveticus* CNRZ 32	Lys-*p*-NA	97	-	-	-	metallo	118
AMP VI	*Lb. delbrueckii* ssp. *bulgaricus* B14	Lys-*p*-NA	95	7	50	1	metallo	119
AMP VII	*Lb. helveticus* LME-511	Lys-*p*-NA	92	7	37	1	metallo	120
AMP VIII	*Lb. casei* ssp. *casei* LLG	Lys/Arg-*p*-NA	87	7	39	1	metallo	121
AMP IX	*Lb. delbrueckii* ssp. *bulgaricus* ACA-DC233	Lys-*p*-NA	98	6	40	1	metallo	122
XAP[6] I	*Lb. delbrueckii* ssp. *lactis*	X-Pro-*p*-NA	165	7	50-55	2	serine	123
XAP IV	*Lb. helveticus* CNRZ 32	X-Pro-*p*-NA	72	7	40	1	serine	124
XAP V	*Lb. delbrueckii* ssp. *bulgaricus* CNRZ 397	X-Pro-*p*-NA	82	7	50	-	serine	125
XAP VI	*Lb. delbrueckii* ssp. *bulgaricus* B14	X-Pro-*p*-NA	170-200	6.5	45	2	serine	126
XAP VII	*Lb. acidophilus* 357	X-Pro-*p*-NA	170-200	6.5	45	2	serine	126
XAP VIII	*Lb. delbrueckii* ssp. *bulgaricus* LBU-147	X-Pro-*p*-NA	270	6.5	50	3	serine	127
PepC-like	*Lb. delbrueckii* ssp. *bulgaricus* B14	-	54	6.5-7	50	1	thiol	128

4. AMP, general amino peptidase. 6. XAP, X-prolyl-dipeptidy aminopeptidase.

Table 1. Cont'd.

Peptidase	Strain	Substrate	MW/kDa	Opt. Activity pH	Opt. Activity °C	Subunits	Class	Reference
Di-/Tripeptidases								
Lactococcus								
DIP[8]		dipeptides	25 and 34	7	30		metallo	129
DIP I	*Lc. bv. diacetylactis* CNRZ 267	dipeptides	51	7.5	-	1	metallo	130
DIP II	*Lc. lactis* ssp. *cremoris* H61	dipeptides	100	8	-	-	metallo	131
DIP III	*Lc. lactis* ssp. *cremoris* Wg2	dipeptides	49	8	50	1	metallo	132
TRP[9] I	*Lc. bv. diacetylactis* CNRZ 267	tripeptides	75	7	35	-	metallo	104
TRP II	*Lc. lactis* ssp. *cremoris* Wg2	tripeptides	103-105	7.5	55	2	metallo	133
TRP III	*Lc. lactis* ssp. *cremoris* AM2	tripeptides	105	8.6	-	2	metallo	134
PRD[10]	*Lc. lactis* ssp. *cremoris* H61	X-Pro dipeptides	43	6.5-7.5	-	-	metallo	135
PRD	*Lc. lactis* ssp. *cremoris* AM2	X-Pro dipeptides	42	7.35-9.0	-	-	metallo	136
PIP[11]	*Lc. lactis* ssp. *cremoris* HP	Pro-X-(Y) peptides	100	8.5, 37	-	2	metallo	98
Lactobacillus								
DIP IV	*Lb. delbrueckii* ssp. *bulgaricus* B14	dipeptides	51	7	50	1	metallo	137

8. DIP, dipeptidase. 9. TRP, trippetidase. 10. PRD, prolidase. 11. PIP, proline aminopeptidase.

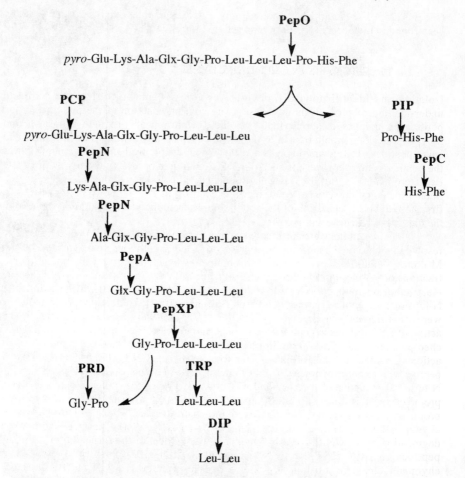

Figure 7 Degradation of a hypothetical oligopeptide by the combined action of known lactococcal peptidases.

Penicillium spp. (mould-ripened varieties), *Br. linens* (smear-ripened varieties), and *Propionibacterium* spp. (Swiss varieties), see review by Fox et al.[91] Yeasts and moulds can grow on the surface of many soft cheeses (principally of the genera *Kluveromyces*, *Debaryomyces* and *Saccharomyces*) and these microorganisms have, primarily, intracellular proteolytic activity (see ref. 145). *Geotrichum candidum* can grow on the surface of Camembert made from raw milk and produces extra- and intracellular proteinases.[146]

Despite the findings by a number of authors (see McSweeney et al., ref. 26) that the NSLAB can dominate the microflora of Cheddar-type cheese during much of its ripening, the proteolytic system of these organisms has received little attention compared with that of *Lactococcus*. Mesophilic lactobacilli (the group to which the majority of NSLAB appear to belong), possess cell wall-associated and intracellular proteinases but their specificity on the caseins has not been determined. A range of peptidases, including intracellular exopeptidases, dipeptidases, aminopeptidases and endopeptidases, have been identified in *Lactobacillus, Micrococcus and Pediococcus* (see refs. 25, 91, 147 and 148). Interestingly, carboxypeptidase activity, which has not been found in lactococci, has been reported in *Lb. casei*.[148]

4. PROTEOLYSIS IN CHEDDAR CHEESE

Isolation and identification of individual peptides is fundamental to the complete understanding of proteolysis in cheese. The combined action of the proteinases and peptidases discussed above leads to the formation of products ranging from polypeptides comparable in size to the intact caseins, through intermediate-sized and small peptides, to free amino acids and their degradation products. The complexity of the peptide system in cheese is such that fractionation is necessary to fully demonstrate its heterogeneity. Various fractionation schemes have been proposed.[2, 142, 149-153] Such a fractionation scheme, modified from Singh et al.,[142] is shown in Figure 8. The first step involves separation of the water-soluble from the water-insoluble peptides by the method of Kuchroo and Fox.[154]

Peptides in the water-insoluble fraction can be visualized by PAGE and partially resolved by anion exchange chromatography on DEAE-cellulose in the presence of 4.5 M urea.[45] Individual water-insoluble peptides were isolated from chromatographic fractions of a 3 month-old Cheddar cheese made with *Lc. lactis* ssp. *lactis* SK11, by electroblotting onto polyvinylidine difluoride membranes and identified from their N-terminal sequence (Figure 9). Three peptides with slow electrophoretic mobility were identified as γ-caseins (β-CN f29-209, f106-209 and f108-209) formed by the action of plasmin; these peptides are present in the electrophoretograms of a number of cheese varieties. Chymosin has limited action on β-casein in Cheddar, although some action is evident by the formation of a low level of β-CN f1-198/192 (β-I-CN). A peptide with an electrophoretic mobility intermediate between α_{s1}- and β-caseins had the N-terminal sequence of β-casein and was probably one of the proteose peptones, also produced by plasmin. The other peptides identified by McSweeney et al.[45] were produced from α_{s1}-casein by the coagulant. The primary product of the action of chymosin on α_{s1}-casein, α_{s1}-CN f24-199 (α_{s1}-I-CN), was clearly evident. The degradation of α_{s1}-CN f24-199 by chymosin was evident in the formation of 3 further peptides (α_{s1}-CN f24-?, α_{s1}-CN f33-? and α_{s1}-CN f102-?) which correspond to chymosin cleavage sites in α_{s1}-casein.[39] The peptide α_{s1}-CN f110-? does not correspond to the specificities of chymosin, plasmin or the cell wall-associated proteinase of *Lc. lactis* ssp. *cremoris* SK11. This peptide is not produced in cheese made with recombinant chymosin but was identified in cheese made using porcine pepsin (Lane and McSweeney, unpublished), suggesting that this peptide is formed by pepsin present in the coagulant. The enzyme responsible for the formation of only one major peptide (α_{s1}-CN f60-?) in the water-insoluble fraction of Cheddar remains unidentified. These results confirm the hypothesis that the primary proteolysis of caseins in Cheddar cheese occurs through the action of chymosin and plasmin and that the microflora of the cheese contributes little at this level of proteolysis.

The water-soluble fraction is first fractionated by diafiltration (DF) through 10 kDa membranes.[151] The DF retentate contains several peptides detectable by PAGE (Figure 10) which have been partially purified by chromatography on DEAE-cellulose (Figure 11). The slower-migrating peptides have been isolated by electroblotting and partially sequenced (Figure 10). The faster-migrating peptides are difficult to isolate by electroblotting and we are attempting to isolate these by preparative HPLC on C_8 and C_{18} columns. In addition to β-lactoglobulin, 9 peptides have been identified, all of which originated from the N-terminal region of β-casein (Figure 10; Singh and Fox, unpublished). Three of the peptides originated at Arg_1, i.e the N-terminus of β-casein and one at Asn_7. These peptides could be formed by the action of microbial enzymes on intact β-casein, or, more likely, on a proteose peptone, e.g. β-CN f1-105/107 (formed by plasmin); the bond Glu_5-Leu_6 is cleaved by the cell wall proteinase (P_{III}-type) of *Lc. lactis* ssp. *cremoris* AM_1.[86] One peptide orginated at Lys_{29} (the primary site of

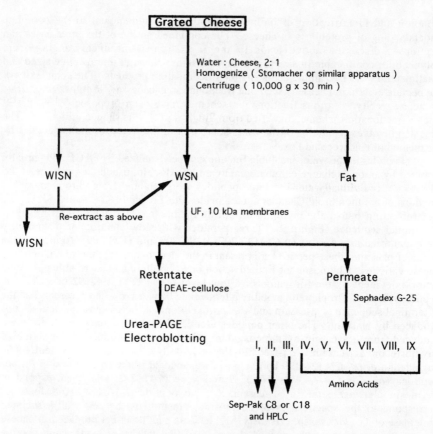

Figure 8   Fractionation scheme for cheese nitrogen.

plasmin action on β-casein). Three of the remaining peptides (β-CN f53-?, 57-? and 58-?) originate at cleavage sites of the cell wall-associated proteinase of *Lactococcus* (see above). The peptide β-CN f58-72 was also isolated from the water-soluble fraction of Cheddar by Stepaniak *et al.*[155] and found to inhibit intracellular lactococcal oligoendopeptidase (LEP).

The DF permeate, containing small peptides and amino acids, is further fractionated by gel permeation chromatography on Sephadex G-25 (Figure 12) and analysed by RP-HPLC (Figure 13). Fraction IV contained most of the free amino acids but no peptides. Fractions V, VI and IX contained Phe, Tyr and Trp, respectively, which adsorb on Sephadex-type gel filtration media. Fractions I to III contained many peptides, some of which have been isolated by chromatography on

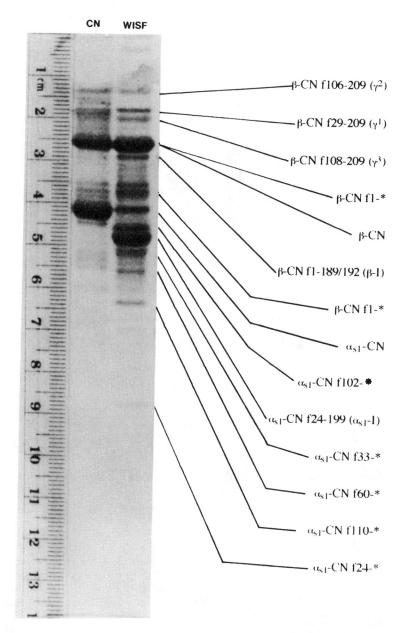

Figure 9 Urea-polyacrylamide gel electrophoretograms (12.5% T, 4% C, pH 8.9) of casein (CN) and water-insoluble (WISF) fraction of 3 month-old Cheddar cheese made with *Lactococcus lactis* subsp. *cremoris* SK 11 indicating the positions of the caseins and the N-terminal of the principal peptides identified.[45] (* Undetermined C-terminus)

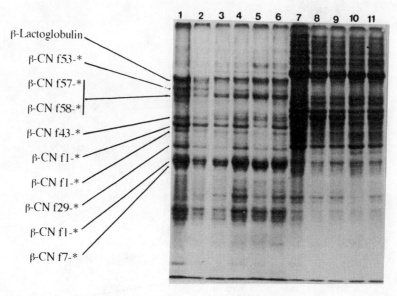

Figure 10   Urea-polyacrylamide gel electrophoretograms (12.5% T, 4% C, pH 8.9) of a DF retentate of a water-soluble extract of Cheddar cheese (lane 1), water soluble extracts from Cheddar (lanes 2 to 6) and of the corresponding cheeses (lanes 7 to 11)

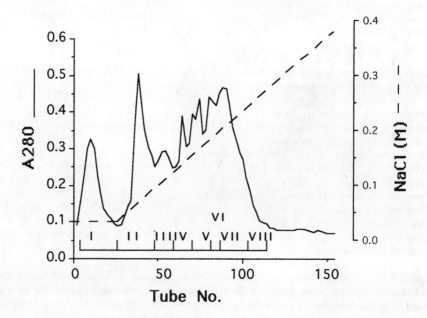

Figure 11   Chromatogram of the DF retentate from a water-soluble extract of Cheddar cheese on DEAE cellulose using a linear NaCl gradient, 0 to 0.5 M, in 50 mM tris-HCl buffer, pH 8.6.

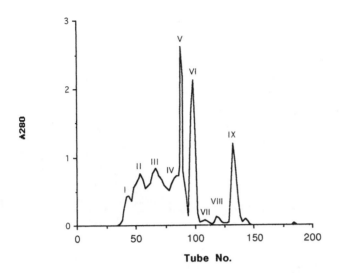

Figure 12 Gel permeation chromatogram of peptides in a 10 kDa ultrafiltration permeate of a water-soluble extract of a mature Cheddar cheese. Freeze-dried extract was dissolved in water and applied to a column (80 x 2 cm) of Sephadex G-25 which was eluted with water (flow rate 0.7 ml min^{-1}); eluate was collected in 3.5 ml fractions and the absorbance at 280 nm determined Fractions were pooled, as indicated, for further analysis.

Sep-Pak® C_{18} reverse phase cartridges (Waters, Milford, MA, USA) and RP-HPLC on C_8 or C_{18} columns using acetonitrile gradients.

The HPLC profile of fraction I of the permeate from Sephadex G25 is shown in Figure 14a. Seven peptides (Figure 14a, 1-7) have been isolated and partially sequenced; their identity must be completed by mass spectrometry. The HPLC profile of fraction II of the permeate from Sephadex G25 is very complex (Figure 14b). Rechromatography of this fraction on Sephadex G25 gave 3 partially resolved peaks (Figure 15), the HPLC profiles of which differed markedly (Figure 16). Some of the peptides (Figure 14b, 1-8) in fractions DFP-G25-II-1 and 2 have been isolated and partially sequenced (Figure 16); additional peptides have been isolated but not yet sequenced. Some of the peptides in fraction DFP-G25-II-3 have also been isolated but we have not yet been able to sequence them, presumably because they are very short. Fraction DFP-G25-III is dominated by 1 peptide (Figure 13) which has been isolated but not yet sequenced.

With the exception of β-CN f7-? and f10-?, all peptides isolated from the DF permeate originated from α_{s1}-casein. These peptides generally corresponded to chymosin cleavage sites (Figure 17). Five peptides (α_{s1}-CN f1-9, f1-13, f11-?, f1-? and f1-?) were probably produced by the action of the lactococcal cell wall proteinase on the peptide α_{s1}-CN f1-23 which is rapidly produced by chymosin (ref. 142 and unpublished). The peptide α_{s1}-CN f11-? may have been produced by cleavage of the bond Gln_9-Gly_{10} by the lactococcal proteinase, followed by the removal of Gly_{10} by an

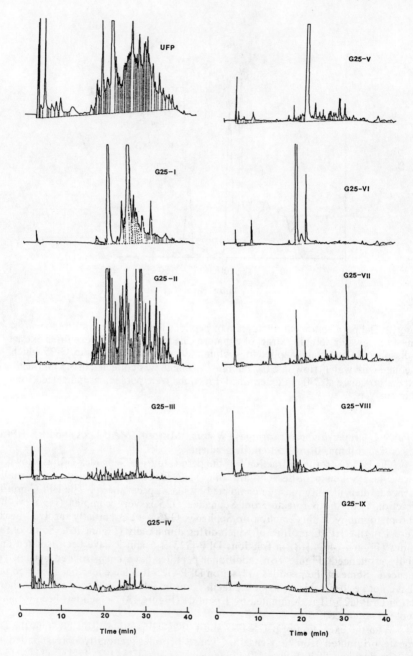

Figure 13 Reversed-phase (C_{18}) HPLC of 10 kDa ultrafiltration permeate of a water-soluble extract of Cheddar cheese and fractions thereof obtained by gel permeation chromatography on Sephadex G-25 (Figure 12).

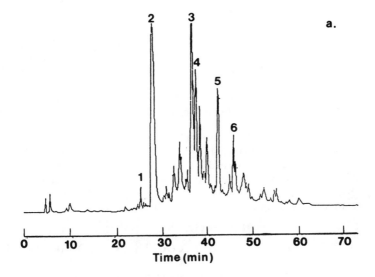

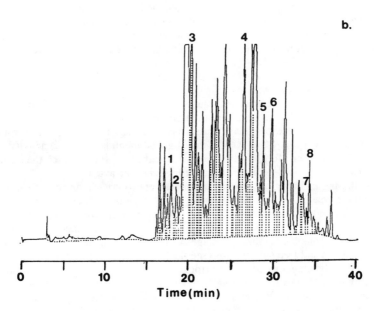

Figure 14   Reversed-phase (C_{18}) HPLC of fractions I and II obtained by gel permeation chromatography of a 10 kDa diafiltration permeate of a water-soluble extract of Cheddar cheese on Sephadex G-25 (Fig. 12). (a) Fraction I, Peak No. 1, α_{s1}-CN f41-*; 2, α_{s1}-CN f1-9; 3, α_{s1}-CN f1-13; 4, α_{s1}-CN f1-*; 5, α_{s1}-CN f1-*; 6, α_{s1}-CN f102-* and β-CN f7-*. (b) Fraction II, Peak No. 1, α_{s1}-CN f44-*; 2, α_{s1}-CN f40-*; 3, α_{s1}-CN f105-*; 4, β-CN f7-*; 5, α_{s1}-CN f11-*; 6, β-CN f10-*; 7, α_{s1}-CN f25-*; 8, α_{s1}-CN f25-*. (* Incomplete sequence.)

aminopeptidase. It is interesting to note that peptides from the N-terminal region of α_{s1}-CN f1-23, particularly f1-9 and f1-13, which are amongst the major peaks in the HPLC elution profile of water-extract of Cheddar, accumulate despite the presence of lactococcal PepX (see above) which should be able to hydrolyse the N-terminus of these peptides (Arg-Pro-Lys...). The peptide α_{s1}-CN f25-? may have been produced by an aminopeptidase acting on a chymosin-produced peptide originating at Phe$_{24}$.

The N-terminal of 2 identified peptides was Ser$_{41}$, suggesting that these originated at the probable chymosin cleavage site Leu$_{40}$-Ser$_{41}$. The peptide with an N-terminal corresponding to Ile$_{44}$ could also have originated from such a peptide after further processing by an aminopeptidase. The peptide corresponding to α_{s1}-CN f102-? also originates at a major chymosin cleavage site (Leu$_{101}$-Lys$_{102}$). The peptide α_{s1}-CN f105-? could have been formed from the former peptide by an aminopeptidase (Singh and Fox, unpublished). Plasmin is also quite active on α_{s1}-casein in this region, cleaving Lys$_{102}$-Lys$_{103}$, Lys$_{103}$-Tyr$_{104}$ and Lys$_{105}$-Val$_{106}$.[71] The enzyme responsible for the formation of the peptide α_{s1}-CN f40-? is not apparent.

5. DEATH AND LYSIS OF *LACTOCOCCUS* IN CHEDDAR CHEESE

In Cheddar and similar cheese varieties, the starter attains maximum numbers at the end of the curd-manufacturing stage. The cells then die at a rate dependent on the strain of *Lactococcus* (typically to 1% of maximum numbers after 3 months). The rate of lysis of the dead *Lactococcus* cells also varies widely with strain. The best information available at present indicates that the only truely external enzyme in the lactococci is the proteinase which is attached to the cell wall where it has ready access to extracellular proteins. The endo- and exopeptidases appear to be intracellular although some of them may be located toward the outside of the cell.[138] Since only hexa- or heptapeptides can be transported into the bacterial cell and since most of the peptides produced from α_{s1}- or β-casein by the cell wall-associated proteinase are larger than this, the lactococcal endo- or exopeptidase must have access to the large casein-derived oligopeptides. However, 80 to 90% of the oligoendopeptidase (PepO) activity appears to be cytoplasmic. Therefore, further work on the mechanism by which the lactococcal cell obtains its amino acid requirements appears to be warranted.

If the intracellular lactococcal peptidases are to contribute to cheese ripening, the cells must lyse or become permeable to quite large molecules and the peptideases must be stable in the cheese environment. Inter-strain differences in the rate of cell lysis appear to be very considerable.[156, 157] Since intracellular lactococcal peptidases are primarily responsible for the final stages of proteolysis in cheese, i.e. the formation of small peptides and free amino acids (see Section 4), fast-lysing starter strains would appear to be advantageous in cheese ripening if these compounds are important in cheese flavour. However, the significance of intracellular exopeptidases remains equivocal. For example, the use of genetically modified starter strains that produced 4 times as much PepN as the wild-type culture, with or without super-production of the cell wall proteinase also, did not accelerate the rate of proteolysis or ripening.[158] As discussed above, α_{s1}-CN f1-9 and f1-13 accumulate in cheese; this is unexpected since PepX of the *Lactococcus* spp. should be able to release the N-terminal Arg-Pro from these peptides, but, so far, no peptide commencing at Lys$_3$ of α_{s1}-casein has been isolated. This might indicate that PepX is not released from the dead cells, that it is unstable or inactive in cheese or that it is unable to hydrolyse these peptides, although PepX from *S. thermophilus* does hydrolyse R.P-*p*NA.[123]

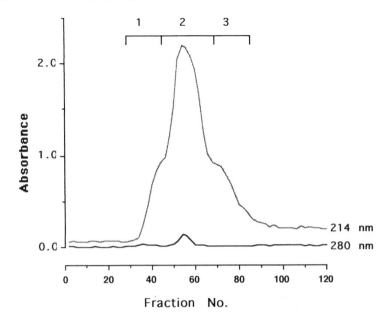

Figure 15 Rechromatography on Sephadex G25 of Fraction II obtained from chromatography of a 10 kDa UF permeate of a water-soluble extract of Cheddar cheese (Figure 12).

Another interesting feature of most of the peptides isolated and characterized so far, both water-soluble and water-insoluble, is that their N-terminal sequence commences at an established cleavage site of chymosin, plasmin or lactcoccal cell wall proteinase. This might suggest that lactococcal intracellular exopeptidases are not very active in cheese.

6. SIGNIFICANCE OF NON-STARTER LACTIC ACID BACTERIA (NSLAB) IN CHEDDAR CHEESE

As discussed above, the starter *Lactococcus* reach maximum numbers in Cheddar at salting and then die off. In contrast, numbers of NSLAB are very low initially (<50 cfu g^{-1} of cheese ex-press) in cheese made in modern factories from a good milk supply in enclosed cheesemaking equipment. However, these grow at a temperature-dependent rate and usually reach ~10^7 cfu g^{-1} after ~3 months.[26] In Irish Cheddar, NSLAB are exclusively mesophilic lactobacilli.[159]

The proteolytic system of mesophilic lactobacilli has not been studed thoroughly, but is probably similar to that of the thermophilic lactobacilli, which appears to be very similar to that of the *Lactococcus*. In Cheddar, the numbers of mesophilic lactobacilli do not decrease significantly during the normal life-time of the cheese (up to 12 months: Fox and Folkertsma, unpublished), suggesting that the cells do not die and lyse, as do the *Lactococcus*. This would suggest that only externally-

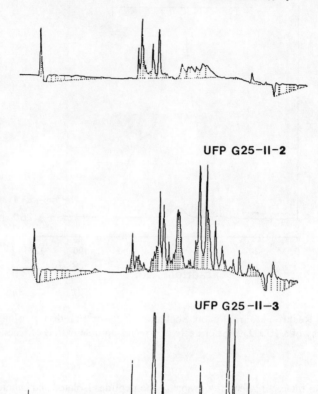

Figure 16 Reversed-phase (C_{18}) HPLC of fractions obtained by gel permeation chromatography on Sephadex G25 of fraction II (Figure 13).

located enzymes of these microorganisms would contribute significantly to proteolysis. The cell wall-associated proteinase of mesophilic lactobacilli has not been studied in detail. However, studies on Cheddar cheese made from raw or pasteurized milk indicate that the indigenous NSLAB in raw milk make a significant qualitative and quantitative contribution to proteolysis and flavour/off-flavour development.[26]

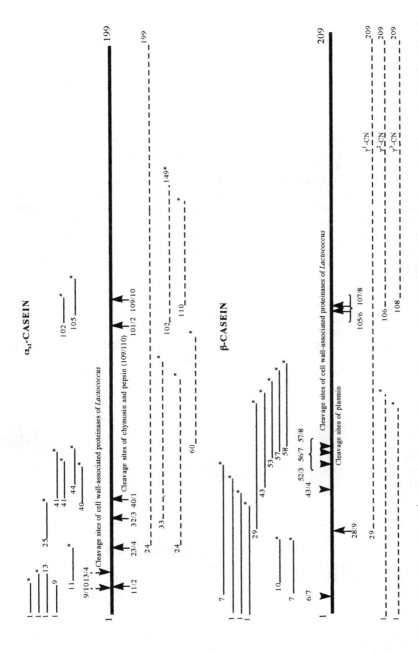

Figure 17 Peptides originating from (a) α_{s1}-casein and (b) β-casein isolated from Cheddar cheese by Singh et al. (ref. 142 and unpublished) and McSweeney et al.[39] Water soluble ———, water insoluble ----------.

Presumably, the NSLAB in pasteurized milk cheese also contribute although probably to a lesser extent, considering the lower numbers normally present initially. However, definitive studies on this appear to be lacking; all the studies on cheese with a controlled microflora have excluded the NSLAB (see ref. 6). A comparative study on cheeses made under aseptic or non-aseptic (open-vat) conditions appears warranted.

7. CONCLUSIONS

Substantial progress has been made on the specificity of the principal proteinases in cheese. Most of the large and several of the small peptides in Cheddar cheese have been isolated and partially identified. The large, water-insoluble peptides are produced either by the action of chymosin on α_{s1}-casein or plasmin on β-casein and the cell wall proteinases of neither the lactococcal starter nor the NSLAB appear to be involved. All the peptides so far identified in the DF retentate of the water-soluble fraction originate from β-casein, apparently through the action of the lactococcal cell wall proteinase. These peptides have not yet been fully identified; they may be retained by the DF membranes because they are sufficiently large or they may be strongly hydrophobic and therefore rejected by the membranes. In contrast, most of the peptides in the DF permeate, which is extremely heterogeneous when analyzed by HPLC, are produced from peptides formed from α_{s1}-casein by chymosin probably by lactococcal cell wall proteinases. The presence of relatively few peptides with an N-terminal sequence that does not commence at a known cleavage site of chymosin, plasmin or lactococcal cell wall proteinase suggests that the intracellular lactococcal exopeptidases are not very active in cheese. Comparative studies on cheeses made with fast-lysing or slow-lysing starter strains should provide useful results on the significance of intracellular lactococcal peptidases. Studies on the combined action of chymosin, plasmin, lactococcal proteinase and possibly NSLAB proteinase(s) in model casein systems should also provide interesting information on proteolysis in cheese.

8. ACKNOWLEDGEMENTS

The financial support of EOLAS, Glasnevin, Dublin, is gratefully acknowledged. The authors also wish to thank Ms Anne Cahalane for assistance in preparing this manuscript.

9. REFERENCES

1. R. Grappin, T.C. Rank and N.F. Olson, J. Dairy Sci., 1985, 68, 531.
2. P.F. Fox, J. Dairy Sci., 1989, 72, 1379.
3. Bulletin 261, 1991, Intern. Dairy Fed., Brussels.
4. P.L.H. McSweeney and P.F. Fox, in 'Cheese: Chemistry, Physics and Microbiology', 2nd edn., P.F. Fox (ed.), Chapman and Hall, London, 1993, Vol. 1, Chapter 9, p. 341.
5. P.F. Fox, P.L.H. McSweeney and T.K. Singh, in 'Chemistry of Structure/Function Relationships in Cheese', E.L. Malin (ed.), Plenum Press, New York, 1994, (in press).
6. P.F. Fox, J. Law, P.L.H. McSweeney and J. Wallace, 1993, 'Cheese: Chemistry, Physics and Microbiology', 2nd. edn., P.F. Fox (ed.), Chapman and Hall, London, 1993, Vol. 1, Chapter 10, p. 389.

7. F.M.W. Visser, Neth. Milk Dairy J., 1976, 30, 41.
8. F.M.W. Visser, Neth. Milk Dairy J., 1977, 31, 120.
9. F.M.W. Visser, Neth. Milk Dairy J., 1977, 31, 188.
10. F.M.W. Visser, Neth. Milk Dairy J., 1977, 31, 210.
11. F.M.W. Visser and A.E.A. de Groot-Mostert, Neth. Milk Dairy J., 1977, 31, 247.
12. B. Reiter, Y. Sorokin, A. Pickering and A.J. Hall, J. Dairy Res.,1969, 36, 65.
13. J.C. Gripon, M.J. Desmazeaud, D. Le Bars and J.L. Bergere, Le Lait, 1975, 55, 502.
14. M.J. Desmazeaud, J.-C. Gripon, D. Le Bars and J.L. Bergere, Lait, 1976, 56, 379.
15. R.B. O'Keeffe, P.F. Fox and C. Daly, 1976, Ir. J. Agric. Res., 15, 151.
16. R.B. O'Keeffe, P.F. Fox and C. Daly, J. Dairy Res., 1976, 43, 97.
17. A.M. O'Keeffe, P.F. Fox and C. Daly, J. Dairy Res., 1978, 45, 465.
18. N.Y. Farkye and P.F. Fox, J. Agric. Food Chem., 1991, 39, 786.
19. S. Kaminogawa. and K. Yamauchi, Agr. Biol. Chem., 1972, 36, 2351.
20. S. Kaminogawa and K. Yamauchi, S. Sasaki and Y.Koga, Proc. 20th Intern. Dairy Congr., Paris, 1978, E, 313.
21. S. Kaminogawa, K. Yamauchi, S. Miyazawa and Y. Koga, J. Dairy Sci., 1980, 63, 701.
22. L.B. Larsen, A. Boisen and T.E. Petersen, FEBS Letts., 1993, 319, 54.
23. P.L.H. McSweeney, P.F. Fox and N.F. Olson, Intern. Dairy J., 1994, in press.
24. A. Noomen, Neth. Milk Dairy J., 1978, 32, 26.
25. S.D. Peterson and R.T. Marshall, J. Dairy Sci., 1990, 73, 1395.
26. P.L.H. McSweeney, P.F. Fox, J.A. Lucey, K.N. Jordon and T.M. Cogan, Intern Dairy J., 1993, 3, 613.
27. J. S. Fish, Nature, 1957, 180, 345.
28. D.G. Dalgleish, 1993, in 'Cheese: Chemistry, Physics and Microbiology', 2nd edn., P.F. Fox (ed.), Chapman and Hall, London, 1993, Vol. 1, Chapter 3, p. 69.
29. J.P. Pelissier, J.C. Mercier and B. Ribadeau-Dumas, Ann. Biol. Anim. Biochem. Biophys., 1974, 14, 343.
30. L.K. Creamer, N.Z.J. Dairy Sci. Technol., 1976, 11, 30.
31. S. Visser and K.J. Slangen, Neth. Milk Dairy J., 1977, 31, 16.
32. C. Carles and B. Ribadeau-Dumas, Biochem., 1984, 23, 6839.
33. D.M. Mulvihill and P.F. Fox, Ir. J. Food Sci. Technol., 1978, 2, 135.
34. R.D. Hill, E. Lahav and D. Givol, J. Dairy Res., 1974, 41, 147.
35. C. Carles and B. Ribadeau-Dumas, FEBS Letts., 1985, 185, 282.
36. L.K. Creamer and N.F. Olson, J. Food Sci., 1982, 47, 631.
37. D.M. Mulvihill and P.F. Fox, J. Dairy Res., 1979, 46, 641.
38. E. Pahakala, A. Pihlanto-Leppälä, M. Laukkanen and V. Antila, Meijeritietleenllinen Aikakauskirja, 1989, 47, 39.
39. P.L.H. McSweeney, N.F. Olson, P.F. Fox, A. Healy and P. Hojrup, J. Dairy Res., 1993, 60, 401.
40. D.M. Mulvihill and P.F. Fox, J. Dairy Res., 1977, 44, 533.
41. B.M. Dunn, M.J. Valler, C.E. Rolph, S.I. Foundling, M. Jimenez and J. Kay, Biochim. Biophys. Acta, 1987, 913, 122.
42. P.L.H. McSweeney, Ph.D. Thesis, National University of Ireland, Cork, 1993.
43. M.L. Green and P.D.M. Foster, J. Dairy Res., 1974, 41, 269.
44. M. O'Sullivan and P.F. Fox, Food Biotechnol., 1991, 5, 19.

45. P.L.H. McSweeney, S. Pochet, P.F. Fox and A. Healy, J. Dairy Res., 1994, (in press).
46. P.F. Fox, J. Dairy Res., 1969, 36, 427.
47. D.M. Mulvihill and P.F. Fox, Milchwissenschaft, 1979, 34, 680.
48. M.L. Green, J. Dairy Res., 1977, 44, 159.
49. J.A. Phelan, Ph.D. Thesis, National University of Ireland, Cork, 1986.
50. P.F. Fox and B.F. Walley, J. Dairy Res., 1971, 38, 165.
51. P.A. O'Leary and P.F. Fox, Ir. J. Agric. Res., 1973, 12, 267.
52. K.A. Al-Mzaien, 1985, Ph.D. Thesis, National University of Ireland, Cork, 1985.
53. M.L. Green, S. Angal, P.A. Lowe and F.A. O. Marston, J. Dairy Res., 1985, 52, 281.
54. C.L. Hicks, J. O'Leary and J. Bucy, J. Dairy Sci., 1988, 71, 1127.
55. V.E. Bines, P. Young and B.A. Law, J. Dairy Res., 1989, 56, 657.
56. G. van den Berg and P.J. de Koning, Neth. Milk Dairy J., 1990, 44, 89.
56. A. van Boven, P.S.T. Tan, and W.N. Konings, Appl. Environ. Microbiol., 1988, 54, 43.
57. M. Nuñez, M. Medina, P. Gaya, A.M. Guillen and M.A. Rodríguez-Marín, J. Dairy Res., 1992, 59, 81.
58. M.K. Harboe, Bulletin 269, Intern. Dairy Fed., Brussels, 1992, p.3.
59. M. Teuber, Bulletin 251, Intern. Dairy Fed., Brussels, 1990, p.3.
60. M.B. Grufferty and P.F. Fox, J. Dairy Res., 1988, 55, 609.
61. P.F. Fox and J. Law, 1991, Food Biotechnol., 1991, 5, 239.
62. W.G. Gordon, M.L. Groves, R. Greenberg, S.B. Jones, E.B. Kalan, R.F. Peterson and R.E. Townend, J. Dairy Sci., 1972, 55, 261.
63. W.N. Eigel, Int. J. Biochem., 1977, 8, 187.
64. W.N. Eigel, Int. J. Biochem., 1981, 13, 1081.
65. A.T. Andrews and E. Alichanidis, J. Dairy Res., 1983, 50, 275.
66. W.N. Eigel, J.E. Butler, C.A. Ernstrom, H.M. Farrell Jr., V.R. Harwalkar, R. Jenness and R.McL. Whitney, J. Dairy Sci., 1984, 67, 1599.
67. P.F. Fox and T.K. Singh, 1994, (unpublished).
68. S. Visser, K.J. Slangen, A.C. Alting and H.J. Vreeman, Milchwissenschaft, 1989, 44, 335.
69. D. LeBars and J.-C. Gripon, 1989, J. Dairy Res., 1989, 56, 817.
70. W.R. Aimutis and W.N. Eigel, 1982, J. Dairy Sci., 1982, 65, 175.
71. P.L. H. McSweeney, N.F. Olson, P.F. Fox, A. Healy and P. Hojrup, Food Biotechnol., 1993, 7, 143.
72. W.N. Eigel, J. Dairy Sci., 1977, 60, 1399.
73. E.H. Reimerdes, J. Dairy Sci., 1983, 66, 1591.
74. P.A. Grieve and B.J. Kitchen, J. Dairy Res., 1985, 52, 101.
75. R.J. Verdi and D.M. Barbano, J. Dairy Sci., 1991, 74, 2077.
76. S. Visser, F.A. Exterkate, C.J. Slangen and G.J.C.M. de Veer, Appl. Environ. Microbiol., 1986, 52, 1162.
77. J. Kok, K.J. Leenhouts, A.J. Haandrikman, A.M. Ledeboer and G. Venema, Appl. Environ. Microbiol., 1988, 54, 231.
78. P. Vos, G. Simons, R.J. Siezen and W.M. de Vos, J. Biol. Chem., 1989, 264, 13579.
79. P. Vos, M. van Asseldonk, F. van Jeveren, R. Siezen, G. Simons and W.M. de Vos, J. Bacteriol., 1989, 171, 2795.
80. J. Kok, FEMS Microbiol. Rev., 1990, 87, 15.
81. F.A. Exterkate, 1990, Appl. Microbiol. Biotechnol., 1990, 33, 401.

82. V. Monnet, D. Le Bars and J.-C. Gripon, FEMS Microbiol. Lett., 1986, 36, 127.
83. V. Monnet, W. Bockelmann, J.-C. Gripon and M. Teuber, Appl. Microbiol. Biotechnol., 1989, 31, 112.
84. V Monnet, J.P. Ley and S. Gonzàlez, Int. J. Biochem., 1992, 24, 707.
85. S. Visser, C.J. Slangen, F.A. Exterkate and G.J.C.M. de Veer, Appl. Microbiol. Biotechnol., 1988, 29, 61.
86. S. Visser, A.J.P.M. Robben and C.J. Slangen, Appl. Microbiol. Biotechnol., 1991, 35, 477.
87. J.R. Reid, G.G. Moore, G.G. Midwinter and G.G. Pritchard, Appl. Microbiol. Biotechnol., 1991, 35, 222.
88. J.R. Reid, K.H. Ng, C.H. Moore, T. Coolbear and G.G. Pritchard, Appl. Microbiol. Biotechnol., 1991, 36, 344.
89. P.S.T. Tan, B. Poolman and W.N. Konings, J. Dairy Res., 1993, 60, 269.
90. S. Visser, G. Hup, F.A. Exterkate and J. Stadhouders, Neth. Milk Dairy J., 1983, 37, 169.
91. P.F. Fox, T.K. Singh and P.L.H. McSweeney, in 'Chemistry of Structure/Function Relationships in Cheese', E.L. Malin (ed.), Plenum Press, New York, 1994, (in press).
92. P.L.H. McSweeney, P.F. Fox and J. Law, Milchwissenschaft, 1993, 48, 319.
93. B.A. Law, J. Gen. Microbiol. 1978, 105, 113.
94. P.S.T. Tan, Ph.D. Thesis, University of Groningen, 1992.
95. V. Monnet, M.P. Chapot-Chartier and J.-C. Gripon, Lait, 1993, 73, 97.
96. S. Kaminogawa, T.R. Yan, N. Azuma and K. Yamauchi, J. Food Sci., 1986, 51, 1253.
97. F.A. Exterkate, A.C. Alting and C.J. Slangen, Biochem. J., 1991, 273, 135.
98. R. Baankreis, Ph.D. Thesis, University of Amsterdam, 1992.
99. L. Stepaniak and P.F. Fox, Proc. 29th Norwegian Biochemical Soc. Meeting, Beito, 21-24 January 1993.
100. T.-R. Yan, N. Azuma, S. Kaminogawa and K. Yamauchi, Appl. Environ. Microbiol., 1987, 53, 2296.
101. T.-R. Yan, N. Azuma, S. Kaminogawa and K. Yamauchi, Eur. J. Biochem., 1987, 163, 259.
102. M.J. Desmazeaud and C. Zevaco, Annal. Biol. Anim. Biochem. Biophys., 1976, 16, 851.
103. P.S.T. Tan, K.M. Pos and W.N. Konings, Appl. Environ. Microbiol., 1991, 57, 3539.
104. M.J. Desmazeaud and C. Zevaco, Milchwissenschaft , 1979, 34, 606.
105. A. Geis, W. Bockelmann and M. Teuber, Appl. Microbiol. Biotechnol., 1985, 23, 79.
106. P.S.T. Tan and W.N. Konings, Appl. Environ. Microbiol., 1990, 56, 526.
107. F.A. Exterkate and G.J.C.M. de Veer, Appl. Environ. Microbiol., 1987, 53, 577.
108. G.W. Niven, J. Gen. Microbiol., 1991, 137, 1207.
109. B. Kiefer-Partch, W. Bockelman, A. Geis and M. Teuber, Appl. Microbiol. Biotechnol., 1989, 31, 75.
110. C. Zevaco, V. Monnet and J.-C. Gripon, J. Appl. Bacteriol., 1990, 68, 357.
111. M, Booth, I. Ní Fhaoláin, P.V. Jennings and G. O'Cuinn, J. Dairy Res., 1990, 57, 89.
112. R.J. Lloyd and G.G. Pritchard, J. Gen. Microbiol., 1991, 137, 49.
113. F.A. Exterkate, J. Bacteriol., 1977, 129, 1281.

114. E. Neviani, C.Y. Boquien, V. Monnet, L. Phan Thanh and J.-C. Gripon, Appl. Environ. Microbiol., 1989, 55, 2308.
115. B. Eggimann and M. Bachmann, Appl. Environ. Microbiol., 1980, 40, 876.
116. E.J. Machuga and D.H. Ives, Biochim. Biophys. Acta, 1984, 789, 26.
117. D. Atlan, P. Laloi and R. Portalier, Appl. Environ. Microbiol., 1989, 55, 1717.
118. N.M. Khalid and E.H. Marth, J. Dairy Sci., 1990, 73, 2669.
119. W. Bockelmann, Y. Schulz and M. Teuber, Intern. Dairy J., 1992, 2, 95.
120. H. Miyakawa, S. Korbayashi, S. Shimamura and M. Tomita, J. Dairy Sci., 1992, 75, 27.
121. G. Arora and B.H. Lee, J. Dairy Sci., 1992, 75, 700.
122. E. Taskalidou, I. Dalezios, M. Georgalaki and G. Kalantzopoulos, J. Dairy Sci., 1993, 76, 2145.
123. J. Meyer and R. Jordi, J. Dairy Sci., 1987, 70, 738.
124. N.M. Khalid and E.H. Marth, J. Dairy Sci., 1990, 73, 3068.
125. D. Atlan, P. Laloi and R. Portalier, Appl. Environ. Microbiol., 1990, 56, 2174.
126. W. Bockelmann, M. Fobker and M. Teuber, Intern. Dairy J., 1991, 1, 51.
127. H. Miyakawa, S. Korbayashi, S. Shimamura and M. Tomita, J. Dairy Sci., 1991, 74, 2375.
128. Y. Wohlrab and W. Bockelmann, Intern. Dairy J., 1993, 3, 685.
129. B.A. Law, J. Appl. Bacteriol., 1979, 46, 455.
130. M.J. Desmazeaud and C. Zevaco, Ann. Biol. Anim. Biochem. Biophys., 1977, 17, 723.
131. I.K. Hwang, S. Kaminogawa and K. Yamauchi, Agric. Biol. Chem., 1981, 45, 159.
132. A. van Boven, P.S.T. Tan and W.N. Konings, Appl. Environ. Microbiol., 1988, 54, 43.
133. B.W. Bosman, P.S.T. Tan and W.N. Konings, 1990, Appl. Environ. Microbiol., 1990, 56, 1839.
134. C.L. Bacon, M. Wilkinson, P.V. Jennings, I. Ni Fhaolain and G. O'Cuinn, Intern. Dairy J., 1993, 3, 163.
135. S. Kaminogawa, T. Ninomiya and K. Yamauchi, J. Dairy Sci., 1984, 67, 2483.
136. M. Booth, P.V. Jennings, I. Ní Fhaoláin and G. O'Cuinn, J. Dairy Res., 1990, 57, 245.
137. Y. Wohlrab and W. Bockelmann, Intern. Dairy J., 1992, 2, 345.
138. P.S.T. Tan, M.P. Chapot-Chartier, K.M. Pos, M. Rosseau, C.Y. Boquien, J.-C. Gripon and W.N. Konings., Appl. Environ. Microbiol., 1992, 58, 285.
140. P.S.T. Tan, T.A.J.M. van Kessel, F.L.M. van de Veerdonk, P.F. Zuurendonk, A.P. Bruins and W.N. Konings, Appl. Environ. Microbiol., 1993, 59, 1430.
141. F.A. Exterkate and A.C. Alting, Intern. Dairy J., 1994, (in press).
142. T.K. Singh, P.F. Fox, P. Hojrup and A. Healy, Intern. Dairy J., 1994, 4, 111.
143. M.C. Broome, D.A. Krause and M.W. Hickey, Aust. J. Dairy Technol., 1990, 45, 67.
144. M.C. Broome, D.A. Krause and M.W. Hickey, Aust. J. Dairy Technol., 1991, 46, 6.
145. J.-C. Gripon, V. Monnet, G. Lambert and M.J. Desmazeaud, in 'Food Enzymology', P.F. Fox (ed.), Elsevier Applied Science Publishers, London, 1991, Vol. 1, Chapter 3, p. 131.
146. M. Guéguen and J. Lenoir, Lait, 1975, 55, 145.
147. N.M. Khalid and E.H. Marth, J. Dairy Sci., 1990, 73, 2669.
148. M. El-Soda, J.L. Bergere and M.J. Desmazeaud, J. Dairy Res., 1978, 45, 519.
149. C.N. Kuchroo and P.F. Fox, Milchwissenschaft, 1983, 38, 389.

150. J.W. Aston and L.K. Creamer, N.Z. J. Dairy Sci. Technol., 1986, 21, 229.
151. M. O'Sullivan and P.F. Fox, J. Dairy Res., 1990, 57, 135.
152. E.D. Breen, M.Sc. Thesis, National University of Ireland, Cork, 1992.
153. A.J. Cliffe, J.D. Marks and F. Mulholland, Intern. Dairy J., 1993, 3, 379.
154. C.N. Kuchroo and P.F. Fox, Milchwissenschaft, 1982, 37, 331.
155. L. Stepaniak, P.F. Fox., T. Sorhaug, K. Sletten and R. Tobiassen, Proc. 7th Euro. Conf. Food Chem., Valencia, 1993, Vol. 1, p. 385.
156. M.G. Wilkinson, T.P. Guinee and P.F. Fox, J. Dairy Res., 1993, (in press).
157. M.-P. Chapot-Chartier, C. Deniel, M. Rousseau, L. Vassal and J.-C. Gripon, Intern. Dairy J., 1994, 4, 251
158. A. M. McGarry, M.Sc. Thesis, National University of Ireland, Cork, 1993.
159. K.N. Jordan and T.M. Cogan, Ir. J. Agr. Food Res., 1993, 32, 47.

Manipulation of Proteolysis in *Lactococcus Lactis*

Alfred J. Haandrikman, Igor Mierau, Jean Law, Kees J. Leenhouts, Jan Kok, and Gerard Venema

DEPARTMENT OF GENETICS, UNIVERSITY OF GRONINGEN, KERKLAAN 30, 9751 NN HAREN, THE NETHERLANDS

1 INTRODUCTION

Lactococci are fastidious organisms. For optimal growth they are dependent on the presence of small peptides and free amino acids in the culture medium. Either because of the absence of various functional biosynthetic genes or regulatory mechanisms, lactococci have strain-dependent amino acid requirements[1]. For growth in chemically defined media, the various *L.lactis* strains require the addition of 4 to 15 different amino acids, either in free form or as part of ingestible peptides. The concentration of free amino acids and peptides in milk is only sufficient to allow growth of up to 25 % of the total cell mass of a normal fully-grown culture[2]. Consequently, fast growth and lactic acid production and, therefore, the use of lactococci in dairy fermentations, depends on their ability to degrade milk proteins. Casein degradation and subsequent utilisation of the degradation products requires a complex proteolytic system consisting of proteinases, peptidases and amino acid- and peptide- carriers. These various components of the lactococcal proteolytic system have been subject of intensive biochemical and genetic research for several years[3-5]. In our group, a set of otherwise isogenic lactococcal strains is being constructed that are either overproducing or deficient for one or more components of the proteolytic system. These strains will be derived from *Lactococcus lactis* MG1363, by means of a food-grade integration system, that allows the exclusion of heterologous DNA and antibiotic resistance markers from a modified strain[6]. Analysis of the different properties of these strains may elucidate the role of proteinases, peptidases and transport systems in casein utilisation by lactococci, as well as their role in flavour development in dairy fermentations.

2 THE PROTEOLYTIC SYSTEM

As outlined above, lactococci depend on the efficient degradation of casein to meet their need for essential amino acids, while growing in milk. Since it is unclear which enzymes and/or uptake systems are indeed involved in this pathway, all proteolytic and peptidolytic enzymes as well as the various peptide- and amino acid- transport systems are considered to be part of the lactococcal proteolytic system. This system is schematically depicted in Fig. 1. Only recently, mainly based on genetic data, it emerged

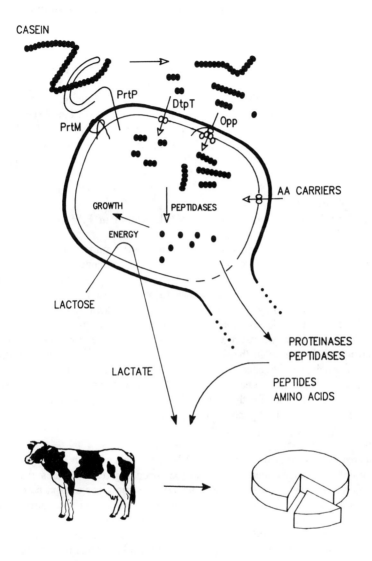

Figure 1   Schematic representation of the proteolytic system of *Lactococcus lactis*. Extracellular casein degradation by the proteinase PrtP is followed by the uptake of amino acids and peptides and subsequent intracellular cleavage of peptides by the various peptidases. Cell lysis may result in the release of peptides, amino acids and peptidases into the culture medium. PrtP, extracellular proteinase; PrtM, proteinase maturation protein; DtpT, di/tripeptide transporter; Opp, oligopeptide transport system; AA, amino acids.

that some of the peptidases are not essential for growth of the organism in milk, although the enzymes may very well contribute to flavour development[7-9].

Extracellular proteolysis
The first step in casein utilisation by lactococci is the extracellular degradation of casein by a cell envelope-associated serine proteinase that, for its activity, depends on the presence of the lipoprotein PrtM[10]. Thus far, this 200-kDa proteinase PrtP is the only lactococcal proteolytic enzyme whose extracellular location is certain. Except for an 93-Kda intracellular proteinase isolated by Muset et al.[11] from L.lactis NCDO763, PrtP is the only lactococcal enzyme capable of hydrolysing casein. All other proteolytic enzymes isolated from lactococci have an almost negligible affinity for casein and hydrolyse only relatively small peptides.

Peptide uptake
The casein degradation products generated by the action of PrtP are taken up by the lactococcal cell.
Separate transport systems for amino acids, a single di-/tripeptide transporter and a single oligopeptide transport system have been described for lactococci[12-14]. Oligopeptides up to at least 8 amino acid residues are internalized by an ATP-driven binding protein-dependent transport system, which was shown to be essential for growth in milk[16]. Di- and tripeptides are taken up by a proton motive force-dependent carrier protein[12,15]. This di-/tripeptide transporter is essential for L.lactis ML3 to grow on casein[12]. This observation indicates that, when grown on casein, at least one essential amino acid has to be taken up by this strain in the form of a di- or tripeptide[12,15]. Considering the fact that as yet no conclusive evidence for an extracellular proteolytic enzyme other than the proteinase PrtP has been presented, it is conceivable that di- and/or tripeptides generated by PrtP from casein provides for the amino acids required for growth. Clear evidence for the production of di- and/or tripeptides by PrtP is still lacking[5].

Peptidases
An ever increasing list of different peptidases is being isolated from *Lactococcus lactis*[4,5]. Ignoring small differences that, of course, exist between the enzymes from different strains of L.lactis, all lactococci seem to produce the same types of peptidases (Table 1). On the basis of their specificity these enzymes can be divided into three groups: (i) proteinases i.e. enzymes that are capable of degrading casein, (ii) (other) endopeptidases, and (iii) aminopeptidases. Until now, no carboxypeptidase activity has been reported in lactococci[4]. As described above, two distinct proteinases have been isolated from L.lactis: PrtP and a 93-kDa intracellular enzyme identified in L.lactis NCDO763[3,11]. The other lactococcal endopeptidases only degrade oligopeptides. Two distinct types of endopeptidases have been purified from lactococci: PepO and LepI. The main difference between these intracellular metallo-endopeptidases concerns their size: PepO is 70 kDa whereas LepI is 98 kDa, and their cleavage specificities: only PepO is capable of cleaving the peptide glucagon[9,17,18]. At least two general aminopeptidases, PepN and PepC, are present in lactococci. The relative amounts of these enzymes seem to vary in different lactococcal strains[19-22]. Recently, a 23-kDa aminopeptidase was purified from L.lactis IMN-C12: peptidase 53[23]. Peptidase 53 appears to be mainly active towards tripeptides carrying an N-terminal Leu-residue, although other tripeptides as well as some di- and tetrapeptides were also hydrolysed[25]. The dipeptidase DIP and the tripeptidase

Table 1: Proteolytic enzymes of *L. lactis*.

Enzyme	Mw (kDa)	Class	Substrate	Leader peptide	Reference
Proteinases					
PrtP	200	serine	casein	yes	3, 38, 43
NisP	54	serine	pre-nisin	yes	45
Neutral proteinase	93	metallo	ß-casein		11
Endopeptidases					
LEPI	98	metallo	α_{s1}-CN(f1-23)		9,18
PepO	70	metallo	Metenkephalin	no	8,16,17
Aminopeptidases					
PepN	95	metallo	Leu/Lys-pNA	no	19,20,50-52
PepC	50	thiol	Leu/Lys-pNA	no	21,22,48
PepA	43	metallo	Glu/Asp-pNA	no	26,37
PepP	45	metallo	Bradykinin		28
PepXP	90	serine	X-Pro-pNA	no	29-32,46,47
PCP	26	serine	Pyr-pNA	no	9,49
Dipeptidase	49	metallo	Leu-Leu		23
PepT	52	metallo	tripeptides	no	24
Peptidase 53	23	metallo	tripeptides		25
Prolidase	43	metallo	X-Pro		33,34
Pro-Iminopeptidase	50	metallo	Pro-X-(Y)		27

PepT can be considered as substrate-size-recognizing general aminopeptidases, only active towards dipeptides and tripeptides, respectively[23,24]. An aminopeptidase also active towards oligopeptides, although less general as compared to PepN and PepC, is PepA. The highest affinity of this enzyme is towards N-terminal Glu- and Asp-residues[26]. The pyrrolidonyl carboxylyl peptidase PCP is a highly specific enzyme that removes N-terminal pyroglutamate-residues from peptides[9]. Because of the high content of proline in casein (11.7 % in α-casein and 16.7 % in ß-casein), enzymes capable of degrading peptides containing proline are thought to be of utmost importance for the degradation of casein by lactic acid bacteria. In lactococci, N-terminal Pro-residues can be removed from di- and tripeptides by the action of the prolyl imino peptidase (PIP)[27]. The enzyme does not depend on the presence of an N-terminal Pro-residue for its activity and may therefore also be regarded as a more general di-/tripeptidase. The tripeptidase PepT from L.cremoris Wg2 also hydrolyses Pro-Gly-Gly[24]. An aminopeptidase, PepP, which preferentially cleaves off N-terminal residues from X-Pro-Pro-Y oligopeptides was recently purified from L.lactis NCDO763[28]. N-terminal prolyl-proline dipeptides, that may appear as the result of the action of PepP, can be removed from oligopeptides by the action of the X-prolyl dipeptidyl aminopeptidase PepXP[29-32]. PepXP is capable of releasing dipeptides from oligopeptides preferentially when the penultimate residue is a Pro. The dipeptides generated in this way can be hydrolysed by the prolidase which is a dipeptidase that specifically degrades dipeptides with a Pro-residue in the second position[33,34].

Peptidase localisation
The location of the lactococcal proteolytic enzymes has been a matter of dispute for a long time. Several aminopeptidases were thought to be extracellular enzymes[3,5,35]. However, a lactococcal peptidase with an undisputable extracellular localization, i.e. an enzyme translocated across the cytoplasmic membrane, has not yet been reported. On the contrary, evidence presented below strongly indicates that all peptidases examined so far are intracellular enzymes. This may also hold true for the recently isolated cell wall peptidase 53[25].
Using antibodies raised against a number of purified lactococcal peptidases in cell fractionation and immunogold-labelling studies, Tan et al. confirmed the intracellular location of PepN, PepC, PepXP, PepT and the endopeptidase PepO[36]. In a separate study, Baankreis studied the location of the peptidases PepA, PepN, PCP, PepXP and both types of endopeptidases[9]. Again, all peptidases were found to be intracellularly located, although, in conflict with cell fractionation studies, immunoelectron microscopy revealed that PepA may be located in the cell envelope[9]. This observation is not supported by genetic data, which indicated that a leader peptide is absent in PepA[37].
Absence of extracellular peptidase activity was indicated by the results of Tynkkynen et al.: proteinase PrtP-deficient L.lactis can not utilise Leu-containing oligopeptides of 4 to 8 amino acids as their sole source of leucine, in the absence of an functional oligopeptide transport system. In these experiments, the L.lactis cells were still capable of using di- and tripeptides. Apparently, the extracellular peptidase activities were so low that insufficient di-/tripeptides were generated to sustain growth of the organism[16]. In an attempt to demonstrate extracellular proteolytic activity in PrtP-deficient lactococcal cells, Baankreis incubated washed cells with the casein fragment α_{s1}-CN(f1-23). All proteolytic activity detected in this experiment, however, could be attributed to the release of intracellular enzymes due to cell lysis[9]. Taken together, these data are in agreement

with the model of the lactococcal proteolytic system as depicted in Figure 1: extracellular hydrolysis of casein is accomplished by the proteinase PrtP alone. Apparently, this hydrolysis produces sufficient di-, tri- and oligopeptides to sustain growth of lactococci in milk. Ultimately, cell lysis may result in the release of peptides, amino acids and peptidases into the culture medium.

3 THE GENES

Proteinase-specifying genes

The first proteolytic enzyme of *L.lactis* to be characterised both biochemically and genetically was the extracelular cell envelope-associated proteinase PrtP[38,39]. PrtP is initially synthesised as a pre-pro-proteinase which for its autoproteolytic maturation, and therefore for its activity, depends on the presence of the extracellular molecular chaperon PrtM[40,41]. In most lactococcal strains, the genes encoding the lipoprotein PrtM (*prtM*) and the proteinase PrtP (*prtP*) reside on a plasmid. In *L.cremoris* SK11 the *prt* genes are flanked by IS-elements, thus forming a composite type transposon[42,43]. The mature active proteinase, devoid of its signal sequence and pro-peptide, remains associated to the cell envelope by means of a membrane anchor in the extreme C-terminus of the molecule. The *prtP* encoded proteinases show considerable similarity with subtilisins, the much smaller serine proteinases secreted by *Bacillus* sp. This similarity especially applies to the regions containing the three amino acids of the subtilisin active site[3,38]. The genetic analysis of the lactococcal proteinase and its maturation, as well as the proteinases from other lactic acid bacteria, have been thoroughly reviewed[3,5,41]. Recently, a second extracellular lactococcal serine proteinase was described: NisP[45]. Like PrtP, NisP is initially synthesised as a pre-pro-proteinase; also the mature NisP remains associated to the lactococcal cell by means of a membrane-anchor. The enzyme has a well defined function in nisin biosynthesis, namely the proteolytic processing of the nisin precursor. The gene specifying NisP is part of the nisin biosynthetic operon, which is only present in nisin-producing lactococci. Although a contribution of NisP to flavour development in dairy fermentations can not be excluded, NisP is not considered to be part of the proteolytic system[45].

Peptidase-specifying genes

The X-prolyl dipeptidyl aminopeptidase PepXP is considered to be an important component of the proteolytic system, due to its capacity to efficiently degrade proline containing peptides. For the cloning of the gene encoding PepXP (*pepXP*) from *L.cremoris* P8-2-47, use was made of the fact that *E.coli* does not contain PepXP activity. A chromosomal DNA bank in *E.coli* was screened for the expression of the *pepXP* gene by a simple plate assay using the chromogenic substrate glycyl-L-prolyl-ß-naphthylamide[46]. In a similar approach, now using a *pepXP*-deficient *L.lactis* as a host for the cloning, the *pepXP* gene from *L.lactis* NCDO 763 was cloned by Nardi *et al*[47]. Comparison between the deduced amino acid sequences of the enzymes from both strains, showed only seven amino acid substitutions[47]. An enzymatic plate assay similar to the one used for the cloning of *pepXP*, in which L-leucyl-ß-naphthylamide was used as a substrate, identified the gene encoding the general aminopeptidase PepC from *L.cremoris* AM2[48]. An *E.coli pepN* mutant, incapable of hydrolysing chromogenic aminopeptidase substrates, was used as a cloning host in this experiment. Similarly, the *pepC* gene from *L.lactis* MG1363, a lactococcal strain producing PepC in small quantities when compared to *L.cremoris* AM2, was picked up using L-alanyl-ß-naphthylamide as a substrate. The aminopeptidase activity produced by *E.coli* carrying the cloned gene could be identified as PepC by using monoclonal antibodies directed towards the purified enzyme[49]. Also, the gene encoding a dipeptidase from *L.lactis* MG1363 was recently cloned by selecting for complementation in the multiple peptidase-negative *E.coli* mutant CM89. This strain is capable of degrading the dipeptide leucyl-methionine when carrying the cloned dipeptidase gene[55].

A gene from *L.lactis* MG1363 encoding the other general aminopeptidase, *pepN*, was cloned using antibodies directed towards the purified enzyme[50,51]. The N-terminal amino acid sequence of purified PepN from *L.cremoris* Wg2 had been published by Tan et al.[19]. Using synthetic oligonucleotides based on this N-terminal amino acid sequence in a PCR reaction, a DNA fragment was synthesised that was subsequently used as a probe for the cloning of *pepN* from *L.cremoris* Wg2[52]. An identical approach was recently used for the cloning of the lactococcal *pepA* gene[37]. For the cloning of the lactococcal *pcp* gene, PCR primers were synthesised on the basis of the known nucleotide sequence of the *pcp* genes from *B.subtilis* and *S.pneumoniae*[44]. The PCR fragment obtained using these primers in a reaction with lactococcal DNA was used as a probe to obtain the complete *pcp* gene of *L.lactis* MG1363[49].

For the cloning of the gene encoding the lactococcal tripeptidase, use was made of the fact that the N-terminal amino acid sequence of the enzyme had been determined. A 53-bp oligonucleotide, synthesised on the basis of this N-terminal amino acid sequence, was used to screen a genomic library of *L.lactis* MG1363 in pUC19. The tripeptidase gene (*pepT*) could be identified in this way. The *pepT* gene, which is part of an operon with unknown function, encodes a 46-kDa protein. The deduced amino acid sequence of the lactococcal enzyme shares 47.7 % identical residues with PepT from *S.typhimurium*[53].

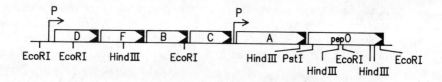

Figure 2   Schematic representation of the chromosomal DNA region from *L.lactis* SSL135, encoding the oligopeptide transport system (*oppDFBCA*) and the endopeptidase (*pepO*). Arrowheads indicate the lengths and orientation of the various genes. The positions of the putative promoter regions (P) are indicated.

The gene encoding the endopeptidase PepO from *L.cremoris* P8-2-47 was isolated from a chromosomal phage bank, using polyclonal antibodies raised against the purified endopeptidase[8]. The amino acid sequence of the endopeptidase, as deduced from the nucleotide sequence, shows a remarkable sequence similarity with eukaryotic neutral endopeptidases (enkaphelinases). Furthermore, nucleotide sequence analysis of the DNA immediately upstream of *pepO* revealed that *pepO* is the last gene in the lactococcal *opp* operon, an operon encoding a binding-protein dependent oligopeptide transport system (Fig.2)[8,16]. The lactococcal *opp* genes were originally cloned in a spontaneous *opp* mutant of *L.lactis* MG1614. This strain is not capable of growing in milk despite the presence of a functional proteinase gene, and *opp* was cloned by selecting for complementation of the growth defect[54]. The oligopeptide transport system consists of two ATP-binding proteins, OppD and OppF, two integral membrane proteins, OppB and OppC, that span the cytoplasmic membrane five to six times, and a substrate binding protein OppA. The lactococcal Opp-system is a member of a large family of so-called ABC (ATP-binding cassette) transporters or traffic ATPases, which includes uptake and secretion systems of both prokaryotic and eukaryotic origin[16]. The substrate binding protein OppA is initially synthesised with a signal sequence typical of prokaryotic prolipoproteins and is, therefore, like the lipoprotein PrtM, expected to be located on the outer surface of the cytoplasmic membrane[16,56]. In fact, PrtP, PrtM, and OppA are the only proteins of the lactococcal proteolytic system that do carry typical N-terminal signal sequences. Comparison of the N-terminal amino acid sequences as deduced from the encoding genes, with the actual N-terminal amino acid sequences of the purified enzymes, revealed that neither PepXP, PepC, PepT, PepN, PepA, nor PepO are posttranslationally processed[8,37,46-53]. The absence of a signal sequence in all of these peptidases fully agrees with the proposed intracellular location of these enzymes.

4 THE GENETIC TOOLS

Changing the proteolytic activity of lactic acid bacteria can be achieved in various ways. An obvious way concerns the introduction into or the removal from the cells of a plasmid harbouring a proteinase gene. This has been accomplished on numerous occasions: lactic acid bacteria losing (by plasmid curing), or acquiring (by conjugation) a plasmid-encoded proteinase[3]. Introduction of new genes, or changing the expression of existing ones, however, requires the availability of a set of genetic tools, such as expression- and integration vectors.

Gene expression
On the basis of the broad-host-range lactococcal plasmid pWV01, a set of cloning vectors has been constructed, including vectors for the isolation of lactococcal promoters[6,57,58]. Exploitation of these promoter screening vectors resulted in the characterisation at the nucleotide level of a number of strong promoters from the lactococcal chromosome[59]. One of these promoters, p32, was used in the construction of the expression vector pMG36e[60]. This vector was successfully applied to the expression in *L.lactis* of both heterologous and homologous genes, including hen egg-white lysozyme, *B.licheniformis* α-amylase, α-galactosidase from guar, and the *B.subtilis* neutral proteinase[60,61]. The latter three gene products are secreted by *L.lactis*.

Chromosomal integration
A plasmid integration system was developed on the basis of the lactococcal plasmid pWV01. For this purpose, a pWV01-derived plasmid was made incapable of replication by deleting the essential *repA* gene, whereas the plus origin of replication initiation, Ori⁺, was retained[6,62,63]. Plasmids lacking the *repA* gene, like plasmid pORI280 in Figure 3, are multiplied in *L.lactis*, *B.subtilis*, or *E.coli* Rep⁺ helper strains in which *repA* is integrated in the chromosome. If the integration vector carries a lactococcal chromosomal DNA fragment and is introduced in a Rep⁻ strain it will integrate via Campbell-type recombination. If the chromosomal piece of DNA is internal to a gene, the integration

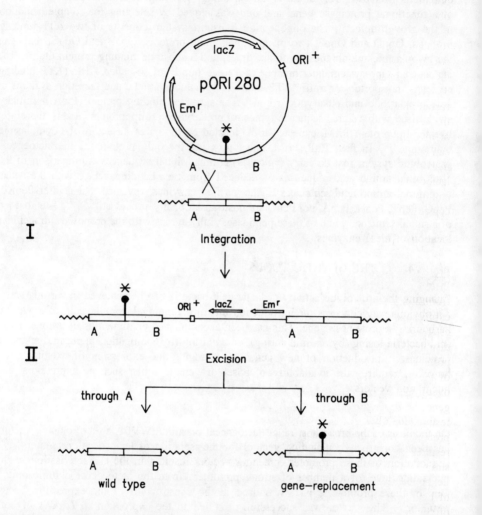

Figure 3: Strategy for obtaining 'food-grade' mutants in *Lactococcus lactis*. Ori⁺, fragment containing the plus origin of replication of pWV01; Em^r, erythromycin resistance marker (For details see text).

will result in disruption of the chromosomal gene[64, 65]. In this way mutations can be generated and heterologous genes can be incorporated in the lactococcal genome. A drawback of this method is the inevitable integration of a selection marker. 'Food-grade' markers are now available to overcome this disadvantage[66]. As an alternative a system for replacement recombination has been developed. By this method a non-replicative plasmid is introduced into the chromosome by homologous recombination involving two cross-over events. If the plasmid contains a chromosomal gene with a mutation the double cross-over produces the equivalent mutation in the chromosome. The strategy for obtaining a mutation in a chromosomal gene, using plasmid pORI280, is outlined in Figure 3. Plasmid pORI280 carries two selection markers: a gene conferring erythromycin resistance, and the *E.coli lacZ* gene under control of promoter p32[6]. When pORI280 containing a chromosomal insert (A + B in Figure 3) with a mutation, is introduced into the strain of interest, the plasmid can integrate either via segment A or B (step I in Figure 3). The transformants will stain blue on agar plates containing X-gal. A second recombination step via the repeats A or B will excise the integrated plasmid rendering the cells *lacZ*⁻ and sensitive to erythromycin and result in either the wild-type situation or in gene replacement (step II in Figure 3). Southern hybridisation, PCR, or bioassays can be used to discriminate between both genotypes. The absence of antibiotic resistance markers in the modified strain offers the possibility of constructing food-grade modified lactococcal strains. Using the above described system in subsequent rounds of mutations, strains carrying multiple chromosomal mutations in, for instance, peptidase genes can be constructed[6].

5 MANIPULATION OF PROTEOLYSIS BY *L.LACTIS*

Proteinase expression

PrtP is the key enzyme in the proteolytic cascade that generates peptides and amino acids from casein. Because of its key function, several groups have attempted to enhance the expression of the *prtP* and *prtM* genes. Although enhanced proteolysis could indeed be established in lactococci, its effect on the growth of the organism is still a matter of dispute[67-70]. The most straightforward approach to obtain a strain with enhanced proteolysis was to clone the *prt* genes on a plasmid with a copy number higher than that of the original proteinase plasmid[3]. The proteinase of *L.cremoris* SK11 was threefold overproduced in *L.lactis* MG1363 upon a tenfold increase of the copy number of the *prt* genes. This overproduction of proteinase, which was shown to be strain dependent, resulted in an increased specific growth and acidification rate in milk[67]. Cloning of the proteinase of *L.cremoris* UC317 in a proteinase-deficient derivative of that strain also resulted in threefold overproduction of the proteinase, when compared to the wild-type strain. However, a beneficial effect for the lactococcal cells harbouring the cloned gene, i.e. an enhanced growth rate in milk, was not observed in these studies[68]. Also, the results of Leenhouts *et al.* indicated that enhancing proteolysis by increasing the copy-number of the *prt* genes in *L.lactis* MG1363 does not result in a higher growth rate of the organism[69]. *L.lactis* strains with 2-3 and 8-9 copies of the *prt* genes were obtained by Campbell-type integration into the chromosome *L.lactis* MG1363 of a plasmid carrying the *prt* genes[69]. Van der Vossen *et al.* placed the originally oppositely orientated *prt* genes in tandem in an operon-like structure, under the control of the strong lactococcal promoter p32[70]. The proteolytic activity of *L.lactis* MG1363 carrying this operon on a plasmid was fivefold increased as compared to that of a strain carrying both genes on a plasmid under control of their natural expression signals[70]. Although its beneficial effect for the producing organism may be disputable, increased proteolysis may still offer a possibility of enhancing/altering flavour production and/or ripening of specific fermented dairy products[71].

In this respect, lactococcal strains producing heterologous proteinases may also offer an interesting possibility. The *B.subtilis nprE* gene, encoding the neutral proteinase, is expressed in *L.lactis* when placed under the control of a lactococcal promoter in the expression vector pMG36e[60]. The enzyme, which is initially synthesised as a pre-pro-proteinase is processed to the active form, and subsequently secreted by *L.lactis*. Similarly, the *B.subtilis* alkaline proteinase or subtilisin, is secreted by *L.lactis* carrying the encoding *aprE* gene cloned in pMG36e[72]. Using a strategy similar to the one illustrated in Figure 3, a food grade *L.lactis* strain has now been constructed that produces active subtilisin. To this purpose, the *B.subtilis aprE* gene placed under control of promoter p32 was cloned in pINT29, an integration vector harbouring the lactococcal *pepXP* gene interrupted by a multiple cloning site[73]. PepXP deficient transformants in which the subtilisin gene was inserted in the chromosomal *pepXP* gene by two subsequent cross-overs as described above, could be selected in a simple plate assay using the chromogenic substrate Glycyl-prolyl-ß-naphthylamide.

Proteinase specificity and stability

All lactococcal proteinases are able to degrade ß-casein with a marked difference in specificity between the P_I- and P_{III}-type proteinases. Only the P_{III}-type proteinases readily degrade α_{s1}- and K-casein[5,74]. Contrary to their differences in specificity of casein breakdown, the lactococcal proteinases show an extremely high degree of conservation: the proteinases from the *L.cremoris* strains Wg2 and SK11, which are representatives of the P_I- and P_{III}-types, respectively, differ in only 44 out of 1902 amino acid residues[75]. Hybrid proteinases have been constructed on the basis of the cloned proteinase genes from *L.cremoris* strains Wg2 and SK11[76]. In this way, two proteinase regions were identified that accounted for the observed differences in specificity. One of these regions contained 7 amino acid differences and corresponds to the subtilisin substrate binding region. The other region contained 8 amino acid differences and is located in the C-terminal part of the proteinase. This region is absent in the subtilisins. Apparently, lactococcal proteinases contain an additional region involved in substrate binding. In this way, various *L.lactis* strains were constructed, some of which produced proteinases with new proteolytic properties, clearly different from those of the parental strains[76]. Recently, Bruinenberg *et al.* demonstrated that altering specific residues in the substrate binding region of the *L.cremoris* SK11 proteinase also leads to the production of proteinases with altered specificity and caseinolytic properties. Interestingly, some of these proteinases appeared to be more stable towards autoproteolytic degradation as compared to the wild-type enzyme[77].

Conclusion: on the function of peptidases

The role of the various components of the lactococcal proteolytic system in the utilisation of casein as a source of essential amino acids is still poorly understood. As outlined above, even the role of the extracellular proteinase PrtP is not completely clear: is the enzyme indeed capable of generating di-/tripeptides from casein? Elucidation of the role of the various peptidases and transport systems in casein degradation is facilitated by the availability of their genes and the genetic tools that allow the construction of well defined *L.lactis* peptidase mutants.

Mayo *et al.* described the specific inactivation of the *pepXP* gene in the chromosome of *L.lactis* MG1363[78]. This mutant strain was obtained by integration, via a double cross-over, of the *pepXP* gene interrupted by an erythromycin resistance marker. Since growth in milk of the mutant relative to the parental strain was not retarded, it was concluded that PepXP is not essential for growth in that medium. Following incubation of the pentapeptide metenkephalin with cell free extracts obtained from wild-type lactococci and

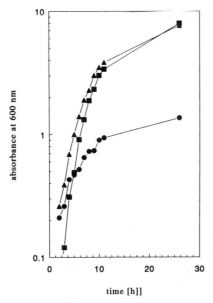

Figure 4: Effect of a *pepO* mutation on growth of *L.lactis* MG1363 on chemically defined medium with casein as the sole source of essential amino acids. ●, MG1363 (proteinase negative); ▲, MG1363(pLP712) wildtype; ■, MG1363 *pepO*⁻(pLP712).

pepXP⁻ mutants, metenkephalin degradation patterns were compared. Disruption of *pepXP* did not impair efficient degradation of metenkephalin, but resulted in a markedly different peptide profile. Not only was the X-prolyl dipeptidyl aminopeptidase activity eliminated upon disruption of *pepXP*, but also the general dipeptidyl aminopeptidase activity. These results suggest that an altered expression of the *pepXP* gene may have a dramatic impact on the peptide composition of fermented milk products[78]. The importance of PepXP in flavour development was substantiated in cheese making experiments: in cheeses manufactured with a *pepXP*-deficient mutant, a generally poor flavour developed[9].

To assess the role of the endopeptidase PepO in casein degradation, a *pepO* mutant was made by gene disruption: the chromosomal *pepO* gene of *L.lactis* MG1363 was interrupted by an erythromycin resistance marker[8]. Growth of this mutant in milk or a chemically defined medium with casein was not affected (Figure 4). Although the close genetic linkage between the *opp* genes and *pepO*, as illustrated in Figure 2, suggests that the gene products are physiologically linked, the PepO function can, apparently, be taken over by other peptidases. By contrast, growth of an *oppA* mutant in milk was severely inhibited[16].

Recently, a tripeptidase PepT-deficient derivative of *L.lactis* MG1363 was constructed[53]. Similar to the *pepXP* and *pepO* mutants described above, the *pepT* mutant grew normally in milk. Apparently, none of the peptidases PepO, PepXP or PepT is essential for lactococci to grow in milk. These observations raise a number of interesting questions: Are these enzymes involved in the utilisation of casein or do they fulfil an entirely different function in the lactococcal cell ? Are non-essential peptidases essential for flavour development in dairy fermentations ? Can the function of individual peptidases be taken over by peptidases with similar and/or overlapping specificities ? To be able to answer these questions, a set of lactococcal strains is presently being constructed in our lab, carrying multiple mutations in chromosomal peptidase genes.

The availability of the genetic tools for the food grade construction of lactococcal strains with altered levels of expression of proteolytic enzymes, now opens the possibility to critically examine the roles of the various enzymes in dairy fermentations.

REFERENCES

1. A. Chopin, FEMS Microbiol.Rev., 1993, 12, 21
2. T.D. Thomas and O.E. Mills, Neth.Milk Dairy J., 1981, 35, 255.
3. J. Kok, FEMS Microbiol.Rev., 1990, 87, 15.
4. J. Kok, J.Dairy Sci., 1993, 76, 2056.
5. G.G. Pritchard and T. Coolbear, FEMS Microbiol.Rev., 1993, 12, 179.
6. K.J. Leenhouts and G. Venema, 'Plasmids, a practical approach', Oxford, 1993, Chapter 3, p. 65.
7. B. Mayo, J. Kok, W. Bockelmann, A.J. Haandrikman, K.J. Leenhouts, and G. Venema, Appl.Environ.Microbiol., 1993, 59, 2049.
8. I. Mierau, P.S.T. Tan, A.J. Haandrikman, B. Mayo, J. Kok, W. Konings, and G. Venema, J.Bacteriol., 1993, 175, 2087.
9. R. Baankreis, PhD Thesis, University of Amsterdam, The Netherlands, 1992.
10. A.J. Haandrikman, R. Meesters, H. Laan, W.N. Konings, J. Kok, and G. Venema, Appl.Environ.Microbiol., 1991, 57, 1899.
11. G. Muset, V. Monnet, and J.-C. Gripon, J.Dairy Res., 1989, 56, 765.
12. E.J. Smid, A.J.M. Driessen, and W.N. Konings, J.Bacteriol., 1989, 171, 292.
13. E.R.S. Kunji, E.J. Smid, R. Plapp, B. Poolman, and W.N. Konings, J.Bacteriol., 1993, 175, 2052.
14. G.H. Rice, F.H.C. Steward, A.J. Hillier, and G.R. Jargo, J.Dairy Res., 1978, 45, 93.
15. A. Hagting, E.R.S. Kunji, Kees J. Leenhouts, B. Poolman, and W.N. Konings, Submitted for publication.
16. S. Tynkkynen, G. Buist, E. Kunji, J. Kok, B. Poolman, G. Venema, and A.J. Haandrikman, J.Bacteriol., 1993, 175,
17. P.S.T. Tan, K.M. Pos, and W. Konings, Appl.Environ.Microbiol., 1991, 57, 3539.
18. T.-R. Yan, N. Azuma, S. Kaminogawa, and K. Yamauchi, Eur. J. Biochem., 1987, 163, 259.
19. P.S.T. Tan, and W.N. Konings, Appl.Environ.Microbiol., 1990, 56, 526.
20. F.A. Exterkate, M. de Jong, G.J.C.M. de Veer, and R. Baankreis, Appl.Microbiol.Biotechnol., 1992, 37, 46.
21. E. Neviani, C.Y. Boquien, V. Monnet, L. Phan Thanh, and J.-C. Gripon, Appl.Environ.Microbiol., 1989, 55, 2308.
22. C.Y. Boquin, F.Nakache, and A. Paraf, Appl.Environ.Microbiol., 1991, 57, 2211.
23. A. Van Boven, P.S.T. Tan, and W.N. Konings, Appl.Environ.Microbiol., 1988, 54, 43.
24. B.W. Bosman, P.S.T. Tan, and W.N. Konings, Appl. Environ. Microbiol., 1990, 56, 1839.

25. S. Sahlstrom, J. Chrzanowska, and T. Sorhaug, Appl.Environ.Microbiol., 1993, 59, 3076.
26. F.A. Exterkate and G.J.C.M. de Veer, Appl.Environ.Microbiol., 1987, 53, 577.
27. R. Baankreis and F.A. Exterkate, System.Appl.Microbiol., 1991, 14, 317.
28. V. Monnet, I. Mars, and J.-C. Gripon, FEMS Microbiol.Rev., 1993, 12, D25.
29. B. Kiefer-Partsch, W. Bockelmann, A. Geis, and M. Teuber, Appl.Microbiol.Biotechnol., 75.
30. M. Booth, I.N. Fhaolain, P.V. Jennings, and G. O'Cuinn, J.Dairy Res., 1990, 57, 245.
31. C. Zevaco, V. Monnet, and J.-C. Gripon, J.Appl.Bacteriol., 1990, 68, 357.
32. R.J. Lloyd, and G.G. Pritchard, J.Gen.Microbiol., 1991, 137, 49.
33. S. Kaminogawa, N. Azuma, I. Hwang, Y. Suzuki, and K. Yamauchi, Agric.Biol.Chem., 1984, 48, 3035.
34. M. Booth, V. Jennings, I.N. Fhaolain, and G. O'Cuinn, J.Dairy Res., 1990, 57,
35. T.D. Thomas, and G.G. Pritchard, FEMS Microbiol.Rev., 1987, 46, 245.
36. P.S.T. Tan, M.-P. Chapot-Chartier, K.M. Pos, M. Rousseau, C.-I. Boquin, J.-C. Gripon, and W.N. Konings, Appl.Environ.Microbiol., 1992, 58, 285.
37. K.I'Anson, H. Griffin, S. Movahedi, F. Mulholland, and M. Gasson, FEMS Microbiol.Rev., 1993, 12, P74.
38. J. Kok, K.J. Leenhouts, A.J. Haandrikman, A.M. Ledeboer, and G. Venema, Appl.Environ.Microbiol., 1988, 54, 231.
39. J. Kok, D. Hill, A.J. Haandrikman, M.J. de Reuver, H. Laan, and G. Venema, Appl.Environ.Microbiol., 1988, 54, 239.
40. A.J. Haandrikman, J. Kok, H. Laan, S. Soemitro, A.M. Ledeboer, W.N. Konings, and G. Venema, J.Bacteriol., 1989, 171, 2789.
41. P. Vos, M. van Asseldonk, F. van Jeveren, R. Siezen, G. Simons, and W.M. de Vos, J.Bacteriol., 1989, 171, 2795.
42. A.J. Haandrikman, C. van Leeuwen, J. Kok, P.Vos, W.M. de Vos, and G. Venema, Appl.Environ.Microbiol., 1990, 56, 1890.
43. W.M. de Vos, I. Boerrigter, P. Vos, P. Bruinenberg, and R.J. Siezen, 'Genetics and Molecular Biology of Streptococci, Lactococci and Enterococci', Washington, 1991, Chapter 3, p.115.
44. Cleuziat, P., A.Awadé and J. Robert-Baudouy. Mol. Microbiol., 1992, 6, 2051.
45. J.R. Van der Meer, J. Polman, M.M. Beerthuizen, R.J. Siexen, O.P. Kuipers and, W.M. de Vos, J.Bacteriol., 1993, 175, 2578.
46. B. Mayo, J. Kok, K. Venema, W. Bockelmann, M. Teuber, H. Reinke, and G. Venema, Appl. Environ. Microbiol., 1991, 57, 38.
47. M. Nardi, M.-C. Chopin, A. Chopin, M.-M. Cals, and J.-C. Gripon, Appl.Environ.Microbiol., 1991, 57, 45.
48. M.-P. Chapot-Chartier, M. Nardi, M-C. Chopin, A. Chopin, and J-C. Gripon, Appl.Environ.Microbiol., 1993, 59, 330.
49. A.J. Haandrikman, I. Mierau, J. Kok, and G. Venema, 1993, Unpublished results.
50. I.J. Van Alen-Boerrigter, R. Baankreis, and W.M. de Vos, Appl.Environ.Microbiol., 1991, 57, 2555.
51. P.S.T. Tan, I.J. Van Allen-Boerrigter, B. Poolman, R.J. Siezen, W.M. de Vos, and W.N. Konings, FEBS , 1992, 306, 9.
52. P. Stroman, Gene, 1992, 113, 107.

53. I. Mierau, A.J. Haandrikman, P.S.T. Tan, C.J. Leenhouts, J. Kok, W.N. Konings, and G. Venema, 1993, Submitted for publication.
54. A. Von Wright, S. Tynkkynen, and M. Suominen, Appl.Environ.Microbiol., 1987, 53, 185.
55. B. Fayard, Unpublished results.
56. A.J. Haandrikman, J. Kok, and G. Venema, J.Bacteriol., 1991, 173, 4517.
57. J. Kok, 'Genetics and Molecular Biology of Streptococci, Lactococci and Enterococci', Washington, 1991, Chapter 3, p.97.
58. J.M.B.M. Van der Vossen, J. Kok, and G. Venema, Appl.Environ.Microbiol., 1985, 50, 540.
59. J.M.B.M. Van der Vossen, D. van der Lelie, and G. Venema, Appl.Environ.Microbiol. 53, 2452.
60. M. van de Guchte, J. Kodde, J.M.B.M. Van der Vossen, J. Kok and G. Venema, Appl.Environ.Microbiol., 1990, 56, 2606.
61. M. van de Guchte, J. Kok, and G. Venema, FEMS Microbiol.Rev., 1992, 88, 73.
62. K.J. Leenhouts, J. Kok, and G. Venema, Appl.Environ.Microbiol., 1991, 57, 2562.
63. K.J. Leenhouts, B. Tolner, S. Bron, J. Kok, G. Venema, and J.F.M.L. Seegers, Plasmid, 1991, 26, 55.
64. K.J. Leenhouts, J. Kok, and G. Venema, Appl.Environ.Microbiol., 1989, 55, 394.
65. K.J. Leenhouts, J. Kok, and G. Venema, Appl.Environ.Microbiol., 1990, 56, 2726.
66. K.J. Leenhouts, J. Kok, and G. Venema, J.Bacteriol., 1991, 173, 4797.
67. P.G. Bruinenberg, P. Vos, and W.M. de Vos, Appl.Environ.Microbiol., 1992, 58, 78.
68. J. Law, P. Vos, F. Hayes, C. Daly, W. M. de Vos, and G.F. Fitzgerald, J.Gen.Microbiol., 1992, 138, 1.
69. K.J. Leenhouts, J. Gietema, J. Kok, and G. Venema, Appl.Environ.Microbiol., 1991, 57, 2568.
70. J.M.B.M. Van der Vossen, J. Kodde, A.J. Haandrikman, G. Venema, and J. Kok, Appl.Environ.Microbiol., 1992, 58, 3142.
71. J. Law, G.F. Fitzgerald, and C. Daly, J.Dairy Sci, 1993, 75, 1173.
72. K. Leenhouts, and G. Venema, Med.Fac.Landbouww.Univ. Gent, 1992, 57/4b, 2031.
73. A.J. Haandrikman, and H. Karsens, 1993, Unpublished results.
74. S. Visser, F.A. Exterkate, C.J. Slangen, and G.J.C.M. de Veer, Appl.Environ.Microbiol., 1986, 52,1162.
75. P. Vos, R.J. Siezen, G. Simons, and W.M. de Vos, J.Biol. Chem., 1989, 264, 13579.
76. P. Vos, I.J. Boerrigter, G. Buist, A.J. Haandrikman, M. Nijhuis, M.B. de Reuver, R.J. Siezen, G. Venema and J. Kok, Protein Engineering, 1991, 4, 479.
77. P.G. Bruinenberg, P. Vos, F.A. Exterkate, A.C. Alting, W.M. de Vos, and R. Siezen,'Stability and stabilization of enzymes', Amsterdam, 1993, p.231.
78. B. Mayo, J. Kok, W. Bockelmann, A.J. Haandrikman, K.J. Leenhouts, and G. Venema, Appl.Environ.Microbiol., 1993, 59, 2051.

New Starter Cultures for Cheese Ripening

Barry A. Law

INSTITUTE OF FOOD RESEARCH, READING LABORATORY, EARLEY GATE, WHITEKNIGHTS ROAD, READING, RG6 2EF, UK

Introduction

The central function of the acidifying culture (starter lactic acid bacteria) in the maturation of Cheddar cheese was established more than twenty years ago (Review; Law, 1984). Although the secondary (non-starter) microbial flora of cheese influence the final flavour profile to an extent dependent on their origin and ability to grow in cheese, most research on mechanisms of flavour development has concentrated on the biochemistry and genetics of starter bacteria generally, and the genus lactococcus in particular. The important characteristics of starter cultures are the ability to produce lactic acid consistently, their resistance to bacteriophage attack, and their ability to mediate in the development of typical, balanced flavour in the stored cheese. These requirements have produced conflicts within the culture industry because they can become mutually exclusive. This paper will discuss this problem, and the opportunities which have emerged from the science base for solving it.

Starter cultures

The key stage in any milk fermentation is the conversion of lactose in milk to lactic acid by the lactic starter culture *(Figure 1)*. These cultures are supplied to the industry either as undefined mixtures of many strains of the appropriate species or, increasingly, as pure defined strain cultures in the form of predetermined mixtures of a small number (2-4) of the correct organism. Cheeses are made with *Lactococcus lactis* (subsp. *lactis, cremoris* or *diacetylactis*) where curd cooking temperatures are below 40^0C (e.g Cheddar, Edam, soft cheeses) or a mixture of the thermophilic lactobacilli (e.g. *Lactobacillus casei, Lb. helveticus*) and streptococci (*Streptococcus thermophilus*) where the curd temperature is raised $>50^0$C (e.g. Emmental, Gruyére).

The technology of producing consistent, pure cultures of starter lactic acid bacteria has advanced rapidly over the past 20 years with the introduction of frozen or freeze-dried, high biomass $>10^{10}$ cfu ml^{-1} or g^{-1}) cultures which can be added directly to the cheese vat. This technique has, in some large creameries, superseded the more traditional approach involving a sequence of mother culture,

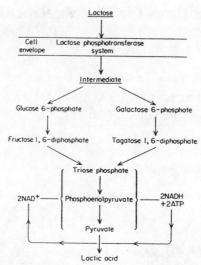

Fig.1. Schematic representation of the pathway for uptake and metabolism of milk lactose to lactic acid in *Lactococcus* spp. used as cheese starter cultures (from Law, 1982).

inoculation culture, bulk starter culture and, finally, vat inoculation (reviewed by Law, 1982). This development represents a success story for the traditional skills of biotechnologists, in that the production of strains that grow to high cell numbers, survive centrifugal harvesting, washing, freezing and drying has been achieved by using selection techniques, and their mass production was made possible by new bioreactor designs which used non-milk media (often produced using enzymic digestion) and highly automated fermentation control, particularly with respect of pH. The prevention of over-acidification of starter cultures, using pumped alkalis, solid neutralizers or even gaseous exchange, had developed into a technology in its own right.

Despite these successes there are still problems associated with the use of lactic starter cultures which can only be solved in the medium- to long-term by the application of the 'new' biotechnologies, involving a detailed knowledge of cell biology, metabolic pathways, enzymology and genetics. The need for such basic understanding of starter bacteria arises from the extraordinary demands now being put on them in cheese factories. Capital investment in automated equipment for cheesemaking is now high, as are raw material costs, and profit margins are low. To cope with this, the larger factories require starter cultures to produce lactic acid more quickly than traditional cultures in less automated units. For example, the time taken to fully coagulate and acidify milk (the 'rennet to mill' time) has been reduced from >5 h to about 3.5 h in many factories. This allows the cheese vats to be used several times a day (often round the clock) but it brings its own problems. The most serious of these is caused by the existence of lytic bacteriophages (phages) which attack the lactic culture and slow down acid production. At best this interferes with the factory schedule, reducing output and profitability, and at worst leads to under-acidified cheese of low quality (and value) which is also susceptible to invasion by pathogens. Most cheese in this latter category has to be discarded or used as a low-value ingredient in processed cheese.

Starter strains which produce acid quickly and are resistant to phages can be obtained by selection, and the problem is generally under control in the industry, but keeping it under control requires heavy technical back-up, either within the industry itself, from Government Institutes, or from the culture supply companies. This back-up takes the form of phage-unrelated strains for use in rotation, phage monitoring to ensure continuity of resistant strains and the supply of starter culture media in which phage replication is inhibited. The prospects for a permanent solution to the phage problem arising from molecular and cell biology are increasing as knowledge of the phenotypic and genetic determinants of natural phage resistance increases.

New starters from new science

Naturally occurring phage-insensitive strains of lactic acid bacteria have been characterized and found to possess multiple defence systems which can operate at different stages of the lytic cycle to prevent adsorption, infection or replication of virulent phages (Klaenhammer, 1989). These defence systems are usually plasmid-encoded and numerous phage-resistance plasmids have now been detected and characterized (Klaenhammer, 1991). Detailed information on the genotypes and phenotypes of phage-resistance plasmids has been derived from phage-insensitive strains which have been naturally selected to operate in culture systems within the industry. It is anticipated that these defined genetic traits will be useful in producing phage-resistant variants which maintain other desirable phenotypic characteristics, such as good acid and flavour production, without interference from uncontrolled genetic events. It is important therefore to understand the nature of the contribution which starter cultures make to flavour development in cheese, in order to achieve this aim.

Much of the evidence derived from the practical industrial use of robust, phage-resistant, efficient acid-producing starters now suggests that they do not produce the fully balanced flavours in traditional hard and semi-hard cheeses that were associated with 'slower' cultures which have not had to operate in the rigour of modern factory environments. There are many possible reasons for this observation, but it is generally assumed to be the result of a combination of (commercial) selection for proteolytic strains (which grow quickly in milk) against peptideolytic strains, and selection for strains which are not easily lysed (by phage or physical stress) so that important intracellular enzymes (including peptidases) are not released into the cheese matrix. It is important to understand that, although not all of the mechanisms through which starter cultures mediate in flavour production are known, there is a strong body of evidence which links cheese taste intensity and quality with the liberation of short peptides and amino acids from casein by the combined action of chymosin, extracellular starter proteinases and intracellular starter peptidases. Failure of the culture to lyse causes an accumulation of bitter peptides and impaired release of short peptides having positive influences on cheese taste (Cliffe *et al.*, 1993), because only the chymosin (coagulant) and extracellular proteinases are free to act on the casein under such circumstances.

Evidence for the importance of particular families of peptides in the development of taste intensity and savoury notes in cheese comes from a variety of sources, not least from the similarity between the short hydrophilic cheese peptides isolated by Cliffe *et al.*, (1993) and the strikingly savoury-tasting peptides in, for example, beef soup (Yamasaki & Maekawa, 1978). There is other, more circumstantial evidence; thus, Cliffe *et al.*, (1989) developed a high-resolution liquid chromatographic method to resolve low molecular weight peptides in ripening cheese and showed that accelerated ripening by peptidase mixtures was characterized by the sequential production and breakdown of hydrophobic (bitter) peptides, to leave, in the best, most intensely flavoured cheese, a preponderance of hydrophilic peptides which molecular weight profiles suggested that they were made up of 2-3 amino acid residues (Cliffe *et al.*, 1993, Fig. 2). One peptide family in particular was present in concentrations that correlated well with enzyme-induced (*not time induced*) flavour intensity (Cliffe & Law, 1991) and this is under investigation at present.

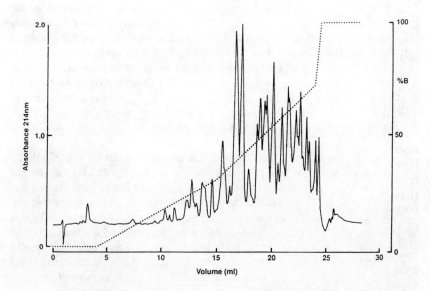

Fig.2. Reverse-phase chromatography of cheese-flavoured, water soluble gel filtration G-25 fraction of high quality Cheddar cheese on a Pharmacia Pep RPC HR 5/5 column. Elution was with Solvent A (0.1%, v/v. TFA in H_2O) and Solvent B (0.1, v/v. TFA in methanol); (Cliffe *et al.*, 1993).

The study of the starter culture-derived enzymes, including those which liberate peptides involved naturally in cheese flavour development, is likely to drive new developments in culture technology for improved and accelerated flavour. This can operate both at the strain selection level, and at the level of genetic engineering. For example, researchers in industrial laboratories have used a cheese slurry method to screen commercial cultures from their vast collection on the basis of their ability to produce low molecular weight N (i.e. nitrogenous materials) in cheese curd and to generate savoury flavour notes (Law 1990). By then selecting *lac*-variants of these cultures, it was possible to show that they could augment the basic, clean, acid flavour produced by the most efficient, phage-resistant homofermentative cheese cultures, with a savoury\cheese flavour.

As our understanding increases of the enzymes and enzyme products that are behind such phenomena, we have the opportunity to isolate, characterize biochemically and clone the enzymes, such that they are expressed more favourably in the most efficient, phage-resistant starters. The most promising enzymes for such cloning are already beginning to emerge. For example, several such enzymes have been purified from strains of *Lac.lactis* subsp. *cremoris*, including an aminopeptidase A (Exterkate and de Veer, 1987), a dipeptidase (Van Boven, et al., 1988), a tripeptidase (Bosman et al, 1990), an X-prolyl dipeptidyl aminopeptidase (Keifer-Partsch et al. 1989; Booth et al. 1990) and aminopeptidases with activity against a wide range of dipeptides and tripeptides (Neviani et al., 1989; Tan and Konings, 1990). Fewer studies have been carried out on the enzymes of *Lac. lactis* subsp. *lactis*, through Kaminogawa, et al., (1983) used cluster analysis to demonstrate that *cremoris* and *lactis* subspecies of *Lac. lactis* formed distinct groups on the basis of peptidase profiles. It is therefore possible that there are significant differences in the characters and types of these enzymes between subspecies which contribute to their different cheesemaking properties. Further studies of the enzymes of *Lac. lactis* subsp. *lactis* are required if an adequate comparison is to be made.

It is clear that the subsp. *lactis* has been somewhat neglected in peptidase research because its strains are not renowned for their flavour-producing abilities. However, this shortcoming appears to be due to their vigorous growth, and slow release of intracellular enzymes, rather than to any inherent lack of peptidases (Law et al., 1974; Chapot-Chartier, et al. (1994). Two peptidases have recently been characterized from this subspecies, an X-Pro dipeptidyl peptidase (Zevaco, et al., 1990; Lloyd and Pritchard, 1991) and an aminopeptidase which shows specific activity against acidic N-terminal residues (Niven, 1991). The substrate specificity of the latter enzyme was tested against a substantial range of aminoacyl-alanine dipeptides and it was shown to be active against N-terminal aspartyl and glutamyl residues. It was also active against tripeptide substrates but did not cleave acidic C-terminal residues. It was therefore designated as an aminopeptidase A (EC 3.4.11.7). In addition, this enzyme also hydrolysed seryl alanine. An aminopeptidase A has previously been purified from *Lac. lactis* subsp. *cremoris* (Exterkate and de Veer, 1987) but activity against seryl residues was not reported, nor has this activity been observed in aminopeptidase A purified from mammalian sources (Benajiba and Maroux, 1980; Danielsen et al. 1980; Tobe et al. 1980). Many of the serine residues in caseins are phosphorylated (Eigel *et al.*, 1984) but it is possible that this enzyme may have activity against phosphorylated serine residues if the acidic side-chain causes them to act as aspartate analogues. This activity may therefore be of particular importance in the breakdown of casein-derived peptides and also in releasing glutamyl residues, which have known savoury properties. Lactic cultures also produce dipeptidyl peptidases which release dipeptides from the N-terminal end of oligopeptide substrates. Such enzymes could be involved in the release of short, flavour-enhancing sequences, and attempts are already under way in a number of European laboratories to clone them for overproduction in commercial starters.

Since most of the peptidase activity of starter cultures is intracellular, attempts to accelerate protein breakdown in cheese have included methods to hasten cell lysis in the early stages of maturation. Techniques range from the use of lysozyme (Law, Castanon and Sharpe, 1976) to the exploitation of phage lysin. The release of phages from the bacterial host cell can involve cell wall degradation by a lysozyme-like enzyme known as phage lysin. ØML3 is a prolate-headed phage that attacks *Lac. lactis* strains and the lysin gene from this phage has been cloned in *E. coli* and its DNA sequence determined (Shearman *et al.*, 1989). The phage lysin has now been expressed in *Lac. lactis* strains, which were found to be unaffected during exponential growth but lysed after reaching the stationary phase in GM17 medium. In milk, the *Lac lactis* subsp. *cremoris* strains that expressed the lysin gene were found to be less viable than controls (Gasson, personal communication). This system may be of potential in accelerating cheese ripening by causing early and increased cell lysis, releasing the intracellular peptidases to act on the milk protein peptides. It is suggested that this autolytic phenotype could also be exploited as a method of containing genetically manipulated *Lac. lactis* strains by routinely inserting the lysin gene in a suicide vector (Shearman, *et al.*, 1989).

Conclusion

The cheese culture industry and the science base are continuing to use advances in cheese chemistry, starter biochemistry and molecular biology to reconcile the conflicting need for efficient acid-producing cultures and the maintenance of traditional taste profiles in cheese. The core research currently involves the identification, isolation and characterization of proteinases and peptidases which collectively can release amino acids and taste-enhancing peptides from caseins. The genetic elements coding for these enzymes, and controlling their production, are also being isolated with the aim of designing and producing new strains of starter cultures, and making available in large quantitites the enzymes themselves, through heterologous gene cloning strategies. This overall approach is summarised in Fig.3.

Biotechnology of Peptides

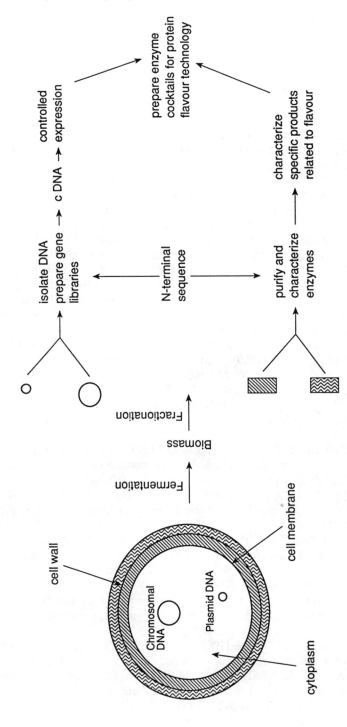

Fig 3. Schematic representation of the core research activities designed to understand and exploit enzymes of *Lactococci* in starter cultures, and cheese ripening technology.

Bibliography

BENAJIBA, A. and MAROUX, S. (1980). Purification and characterization of an aminopeptidase A from hog intestinal brush-border membrane. *European Journal of Biochemistry* **107**, 381-388.

BOOTH, M., NI FHAOLAIN, I., JENNINGS, P.V. and UINN, G. (1990). Purification and characterization of a post-proline dipeptidyl aminopeptidase from *Streptococcus cremoris* AM2. *Journal of Dairy Research.* **57**, 89-99.

BOSMAN, B.W., TAN, P.S.T and KONINGS, W.N. (1990). Purification and characterization of tripeptidase from *Streptococcus cremoris* Wg2. *Applied and Environmental Microbiology* **56**, 1839-1843.

CHAPOT-CHARTIER, M.-P., DENIEL, C., ROUSSAU, M., VASSAL,L. and GRIPON, J.C., (1994). Autolysis of two strains of *Lactoccus lactis* during cheese ripening. *International Dairy Journal.* **4**, 251-269.

CLIFFE, A.J., MARKS, J.D. and MULHOLLAND, F.(1993). Isolation and characterisation of non-volatile flavours from cheese: peptide profile of flavour fractions from Cheddar cheese. *International Dairy Journal.* **42**, 1761-1765.

CLIFFE, A.J., REVELL, D., and LAW, B.A. (1989). A method for the reverse phase HPLC of peptides from Cheddar Cheese. *Food Chemistry* **34**, 147-160.

CLIFFE, A.J., and LAW, B.A. (1991).A time course study of peptide production in accelerated-ripened Cheddar cheese using reverse phase HPLC. *Food Biotechnology* **5**, 1-17.

DANIELSON, E.M., NOREN, O., SJOSTROM, H., INGRAM, J. and KENNEY, A.J. (1980). Proteins of the kidney microvillar membrane. *Biochemical Journal* **189**, 591-603.

EIGEL, W.N., BUTLER, J.E., ERNSTROM, C.A., FARRELL, H.M., HARWALKAR, V.R., JENNESS, R. and WHITNEY, P.M. (1984). Nomenclature of proteins of cow's milk: fifth revision. *Journal of Dairy Science* **67**, 1599-1631.

EXTERKATE, F.A. and VEER, G.J.C.M. (1987). Purification and some properties of a membrane-bound Aminopeptidase A from *Streptococcus cremoris*. *Applied and Environmental Microbiology* **53**, 577-583.

KAMINOGAWA, S., NINOMIYA, T and YAMAUCHI, K. (1983). Aminopeptidase profiles of lactic streptococci. *Journal of Dairy Science* **67**, 2483-2492.

KIEFER-PARTSCH, B., BOCKELMANN, W., GEIS, A. and TEUBER, M. (1989). Purification of an X-prolyl-dipeptidyl aminopeptidase from the cell wall proteolytic system of *Lactococcus lactis* subsp. *cremoris*. *Applied Microbiology and Biotechnology,* **31** 75-78.

KLAENHAMMER, T.R.(1989). Genetic characterization of multiple mechanisms of phage defense from a prototype phage-insensitive strain. *Journal of Dairy Science.* **72**, 3429-3442.

KLAENHAMMER, T.R. (1991). Development of bacteriophage-resistant strains of lactic acid bacteria. *Biochemical Society Transactions.* **19**, 675-681.

LAW, B.A., CASTANON, M. and SHARPE, M.E. (1976). The contribution of starter streptococci to flavour development in Cheddar cheese. *Journal of Dairy Research* **43**, 301-311.

LAW, B.A., SHARPE, M.E. and REITER, B. (1974). The release of intracellular dipeptidase from starter streptococci during Cheddar cheese ripening. *Journal of Dairy Research.* **41**, 137-146.

LAW, B.A., (1982). Cheeses. In *Economic Microbiology Vol.7 Fermented Foods.* (A.H.Rose, Ed.) pp147-198. Academic Press London.

LAW, B.A., (1984). Mechanisms of cheese ripening. In *Advances in the Microbiology and Biochemistry of Cheese and Fermented Milk.* (F.L.Davies and B.A.Law, (Eds). pp 189-208. Elsevier Applied Science, Amsterdam.

LAW, B.A., (1990). The application of biotechnology for accelerated ripening of cheese. *Proceedings of the XXIII International Dairy Congress.* **Vol.2.** pp 1616-1624. Mutual Press, Ottowa.

LLOYD, R.J. and PRITCHARD, G.G. (1991). Characterization of X-prolyl dipeptidyl aminopeptidase from *Lactococcus lactis* subsp. *lactis. Journal of General Microbiology* **137**, 49-55.

NEVIANI, E., BOQUIEN, C.Y., MONNET, V., PHAN THANH, L. and GRIPON, J.-C. (1989). Purification and characterization of an aminopeptidase from *Lactococcus lactis.* subsp.*Cremoris* AM2. *Applied and Environmental Microbiology* **55**. 3208-2314.

NIVEN, G.W. (1991). Purification and characterization of aminopeptidase A from *Lactococcus lactis* subsp. *lactis* NCDO 712. *Journal of General Microbiology* **137,** 1207-1212.

SHEARMAN, C., UNDERWOOD, H., JURY, K. and GASSON, M. (1989). Cloning and DNA sequence analysis of a *Lactococcus* bacteriophage lysin gene. *Molecular and General Genetics* **218**, 214-221.

TAN, P.S.T., and KONINGS, W.N. (1990). Purification and characterization of an aminopeptidase from *Lactococcus lactis* subsp. *cremoris* Wg2. *Applied and Environmental Microbiology* **56**, 526-532.

TOBE, H., KOJIMER, F., AOYAGI, T. and UREZAWA, H. (1980). Purification by affinity chromatography using Amastatin and properties of aminopeptidase A from pig kidney. *Biochemica et Biophysica Acta* **613**, 459-468.

VANBOVEN, A., TAN, P.S.T. and KONINGS, W.N. (1988). Purification and characterization of a dipeptidase from *Streptococcus cremoris* Wg2. *Applied and Environmental Microbiology* **54**, 43-49

YAMASAKI, Y and MAEKAWA, K. (1978). A peptide with a delicious taste. *Agricultural and Biological Chemistry* **42**, 1761-1765.

ZEVACO, C., MONNET, V and GRIPON, J-C. (1990). Intracellular X-prolyl dipeptidyl peptidase from *Lactococcus lactis* ssp. *lactis*: purification and properties. *Journal of Applied Bacteriology* **68**, 357-366.

Engineering Pivotal Proteins for Lactococcal Proteolysis

Willem M. de Vos and Roland J. Siezen

DEPARTMENT OF BIOPHYSICAL CHEMISTRY, NIZO, PO BOX 20, 6710 BA EDE, THE NETHERLANDS

1 INTRODUCTION

Recent years have seen a dramatic increase in the development of technologies that allow the engineering of strains of *Lactococcus lactis* used in industrial milk fermentations[1-3]. These developments have had a great impact on the analysis and engineering of complex metabolic conversions that are characteristic of lactococci used as starter cultures, including lactose fermentation[4-8] and production of antimicrobial peptides such as nisin[9-12]. However, without precedent is the importance of genetic engineering in the study of the cascade of proteolytic reactions that allow lactococci to grow in milk and produce flavor precursors from casein[13-15].

Three types of engineering of the lactococcal proteolytic system can be distinguished that are all based on the availability of characterized genes involved in proteolysis. Firstly, cloned genes can be used to complement mutations that have been obtained by classical mutagenesis or they can be modified to create site-specific mutations that result in the complete elimination of functional enzymes. These powerful approaches have been used to study the importance of peptide and amino acid transport systems and the multitude of peptidases present in lactococci. Secondly, selected key enzymes in the proteolytic system can be overproduced in homologous hosts. This allows not only for studying the biochemical characteristics of those enzymes in more detail but also for analyzing the effect on the growth of lactococci in milk and the properties of the fermented milk product. Finally, the introduction of specific amino acid changes in selected enzymes followed by their (over)production has become feasible. This protein engineering approach allows for a detailed analysis of structure-function relationships and the construction of new enzymes with novel properties.

After a brief summary of the present state of the art on lactococcal proteolysis this review will address the engineering studies performed with two key enzymes in the proteolytic system of lactococci, i.e. the cell envelope-located proteinase that initiates casein degradation, and the intracellular aminopeptidase N that has debittering capacity. In addition, further prospects for the engineering of lactococcal proteolytic enzymes and their casein substrates will be discussed.

2 ESSENTIAL FEATURES OF THE *L.LACTIS* PROTEOLYTIC SYSTEM

Three main steps in proteolysis

Currently, three main steps are distinguished in the cascade of reactions that allow lactococci to degrade the milk protein casein and generate the amino acids required to support their efficient growth in milk. These steps are extracellular degradation, transport and intracellular degradation and are schematically depicted in Fig. 1. Their intrinsic properties have been reviewed recently[15].

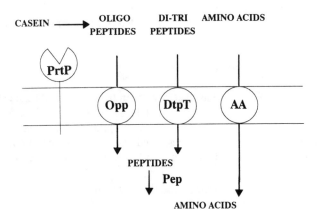

Fig. 1. Schematic representation of the different steps in the cascade of proteolytic degradation of casein. PrtP: cell-envelope located proteinase; Opp: oligopeptide transport system; DtpT: di-tri-peptide transport carrier; AA: dedicated transport systems for various amino acids; Pep: peptidases. For an explanation see text.

A cell-envelope located proteinase is the key enzyme

The first step is the proteolytic degradation of casein by the cell envelope-located proteinase (PrtP in Fig. 1) resulting in the generation of peptides and amino acids outside the lactococcal cell (see below). It is possible that intracellular peptidases released from lysed lactococcal cells also participate in this extracellular degradation but definite proof for the contribution of these peptidases has not yet been provided. Presently, there is no convincing evidence supporting the often-claimed presence of extracellular peptidases, as has recently been discussed[15]. The cell-envelope located proteinase is essential for growth in milk – strains that are deficient in this key proteolytic enzyme have lost the capacity to grow in milk without the addition of peptides[16]. Recently, a second cell-envelope proteinase, NisP, was detected in some lactococcal strains that produce the antimicrobial peptide nisin[17]. However, this serine proteinase is not likely to be involved in caseinolysis because strains deficient in NisP are known to grow fast in milk. In addition, modelling studies indicated that NisP has a very narrow substrate specificity and most likely only degrades its natural substrate, precursor nisin[18].

Dedicated amino acid and peptide transport systems

In the second step the small peptides and amino acids are transported into the lactococcal cell. Presently, three systems for transport have been identified. The oligopeptide transport system (Opp in Fig. 1) that has the capacity to transport peptides of 4-8 residues is an essential component of this transport machinery since its deletion results in lactococci that can not grow in milk even when provided with peptides[19].

The di- and tripeptide transport system (DtpT in Fig. 1) is also indispensable for growth of lactococcal strains in synthetic media containing casein as sole nitrogen source[20]. For the individual amino acids various carriers (AA in Fig. 1) have been identified but no mutants are available yet to determine their role during growth in milk[21].

Panoply of intracellular peptidases

In the final step, the peptides that have been transported into the cell are subject to further degradation catalyzed by a multitude of general and specific intracellular peptidases (Pep in Fig. 1). However, so far no peptidases have been found that are indispensable for growth in milk as is described in a recent review on the discovery of at least 12 different lactococcal peptidases, half of which have sequenced genes[15].

Relation between the proteolytic system and flavor development: importance of PrtP and PepN

It is evident that various steps in the proteolytic system of lactococci are essential for rapid growth in milk and hence the fast acidification of starter cultures. In addition, the enzymes involved in the proteolytic degradation of the milk protein casein generate amino acids and peptides that contribute to the flavor of the fermented dairy product. Among the great challenges in the last decade have been questions as to what exactly determines this flavor and in what way do the lactococcal enzymes participate in the process of flavor generation. Presently, these questions can not be answered completely but a number of experiments have been carried out aimed at elucidating the relative contribution of the various enzymes. These have been done with mutant lactococcal strains that are deficient in a single enzyme of the proteolytic system. Because of the practical relevance, these experiments have been carried out in cheese and flavor development has been scored as a function of maturation time. In this way it has been established that the cell-envelope located proteinase (PrtP) is a major enzyme controlling the generation of cheese flavor[22]. In addition, studies have been performed with strains that are deficient in the general aminopeptidase N (PepN). These experiments have shown that increasing the amount of PepN-deficient strains in a multiple strain starter increases the formation of bitter defects[23]. This is compatible with the capacity of PepN to debitter trypsin-digested casein hydrolysates (see below)[24].

Because of the essential role of PrtP in growth of lactococcal cells in milk and the contribution of PrtP and PepN in generating cheese flavor, we have intensively studied these enzymes, their properties, and their genes, and set out to engineer these as described below.

3 STRUCTURE AND MODELLING OF THE *L.LACTIS* SK11 PROTEINASE

As early as in 1976 the extracellular location of the caseinolytic proteinase in the cell-wall of various *L.lactis* strains could be established, following the detection of its activity by using labelled casein as substrate[25]. Since then, many reports have appeared on the characterization of these large, 110-150 kDa serine proteinases[26].

An important discovery was the observation that the caseinolytic specificity of proteinases from different strains differed considerably[27]. Different classifications of proteinases have been proposed, that are continuously being refined, based on comparing the degradation of casein, casein fragments and synthetic substrates[27-29]. Two main types of specificities can be distinguished: a PI-type specificity found in proteinases that are able to digest ß- and κ-casein and a PIII-type specificity shared by proteinases that digest ß-, κ- and α_{s1}-casein.

The main casein cleavage sites of these two classes of proteinases have been determined and are covered in recent reviews[28,30]. Currently, it is assumed that the peptides generated from caseinolysis contain all the essential amino acids required to support growth in milk[31]. There appears to be a general preference of the PIII-type proteinase to cleave bonds at which negatively charged residues are present at the P_2 or P_3 positions of the substrate, whilst the PI-type proteinase does not cleave such bonds but has a preference for positively charged residues at those positions (see below)[28-31].

Studies aimed at explaining the differences in caseinolytic specificity had to wait until the primary structures of the two classes of proteinases could be deduced from their genes. These genes were identified and sequenced from various strains, including strain Wg2 containing a proteinase that resembles a PI-type[32] and strain SK11 that contains a PIII-type proteinase[33].

Here we will mainly focus on the PIII-type proteinase from the industrial starter strain SK11 that has the broadest cleavage specificity and does not generate as many bitter peptides from casein as the PI-type proteinases[34].

The *prt* genes from the industrial strain *L.lactis* SK11

Lactococcal strains that are able to grow in milk contain proteinase genes that are usually located on plasmids[35]. In the industrial strain *L.lactis* SK11 the 78-kb proteinase plasmid pSK111 has been identified by curing and transfer studies and subsequently mapped[36-38]. Gene banks of pSK111 DNA in *E.coli* were generated using bacteriophage lambda vectors and screened for the expression of proteinase with antibodies raised against the purified enzyme[38]. By subcloning studies in *L.lactis* a region of approximately 10 kb was identified that contained the *prt* genes. The structure and function of this region was further investigated by sequence analysis, deletion mapping and various expression studies (see Fig. 2).

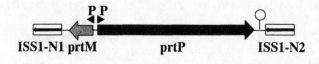

ISS1-N1 prtM   **prtP**   **ISS1-N2**

Fig. 2. Organization of the *L.lactis* SK11 *prt* operon (the bar indicates a size of 1 kb).

The *prt* region is characterized by two divergently transcribed genes, *prtP* and *prtM*, with overlapping promoters located in a region with bidirectional symmetry[39]. The production of the SK11 proteinase appeared to be regulated by the medium composition[40]. Therefore, the regulation of transcription of both genes has been studied and found to be controlled at the transcriptional level by a repression system that is presently the subject of further studies[41].

The *prtP* gene has a size of 5886 bp and encodes the cell envelope-located proteinase PrtP. The *prtM* gene codes for a 33-kD maturation protein that is essential for proteolytic activation of the secreted primary translation product of the *prtP* gene[42]. This maturation protein has been found in all other *prt* gene clusters identified so far and appeared to be a lipoprotein[43] that has homology to a newly recognized family of extracellular chaperons[15]. The SK11 *prt* operon is flanked by two tandem copies of IS*S1*, a widely-distributed lactococcal IS element, and hence has a structure resembling a composite transposon[14,44] (see Fig. 2).

Structure and function of the *L.lactis* SK11 proteinase

The availability of the SK11 *prtP* gene and sequence allowed us to deduce the primary structure of this proteinase with PIII-type specificity (Fig. 3). The proteinase has 1962 residues and starts with a typical *sec*-dependent signal sequence of 33 residues, the functionality of which has been shown in the secretion of foreign proteins from *L.lactis*[39]. Based on the N-terminal sequence of the active proteinase it could be concluded that the SK11 proteinase also contains a 154-residue pro-region and is hence produced as a pre-pro-proteinase[33].

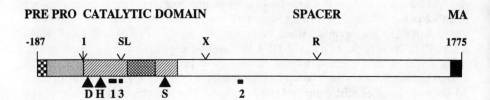

Fig. 3. Structure of the SK11 proteinase in which the different domains and the active site residues are indicated. The 151-residue loop in the catalytic domain is cross-hatched. The presently recognized specificity regions (numbered black bars) and autoprocessing sites (V) are shown below or above the general structure, respectively (1 represents the first residue of the mature proteinase and the main autoprocessing site). MA represents the membrane anchor. Further explanation is provided in the text.

The N-terminal, approximately 500 residues, of the mature proteinase includes the catalytic domain that is homologous to serine proteinases of the subtilisin family, now designated subtilases[45]. This homology not only includes the active site triad residues D30, H94, and S433 but also the substrate binding region and has allowed for modelling the catalytic domain of the SK11 proteinase in order to provide strategies for knowledge-based protein engineering experiments, as will be described below.

The C-terminal end of the SK11 proteinase contains a stop-transfer sequence that may act as a membrane anchor and secures the enzyme in the cell-envelope[32]. This anchor is separated from the catalytic domain by a large spacer region of more than 1000 residues, that shows no homology to proteins with known function. Part of this spacer may be involved in presenting the catalytic domain at the outside of the cell-envelope and it is feasible that sequences preceding the anchor may interact with the cell wall and hence affect the fixation of the proteinase. In addition, protein engineering experiments have shown that the spacer contains a region that is involved in substrate binding and affects caseinolytic specificity (specificity region 2; see Fig. 3 and below).

Modelling the catalytic domain of the SK11 proteinase

Based on the homology with members of the subtilase family for which 3-D structures were available (including subtilisins and thermitase), a model of the catalytic domain of the SK11 proteinase was generated (Fig. 4)[18,45,46,47]. This model illustrates the structurally conserved core typical of subtilases that includes the active site triad residues (D30, H94, and S433). The modelling also identified 10 inserts of 3-151 residues, with a total size of 238 additional residues not present in the core structure of the subtilase family. The extensions vr7 (position 133-135) and vr9 (position 169-172) are located in or close to the substrate binding region of the enzyme, as is the another extension (position 205-219) found to be at the surface of the catalytic domain and here denoted as SL (surface loop, see Fig. 3, Fig. 4, and below). The largest extension (position 238-388) has a size of 151 residues and is sufficiently large to form a separate domain and is indicated as such in Fig. 3 .

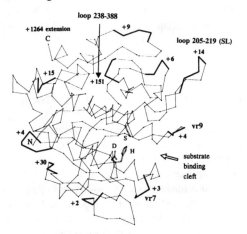

Fig. 4. Model of the catalytic domain of the *L. lactis* SK11 proteinase (for details see text)

Based on the crystal structure of the inhibitor R45-eglin c in complex with subtilisin, various substrates were modelled into the active site of the SK11 proteinase[47]. One of the more relevant substrates is the α_{s1}-casein fragment (1-23) that is generated in cheese by the action of the milk clotting enzyme chymosin. A schematic representation of two α_{s1}-casein fragments with the major cleavage sites by PIII-type (fragment Q-E-V-L▲N-E) and PI-type (fragment I-K-H-Q▲G-L) proteinases is presented in Fig. 5. Electrostatic interactions between the charged residues at position 166 (N in SK11 and D in Wg2) and 138 (K in SK11 and T in Wg2) explain the preferences of the PIII-type or PI-type proteinases for negatively or positively charged residues at the P_3 position of the substrate, respectively. This model can be used to explain the results of some of the proteinase engineering experiments that are described below.

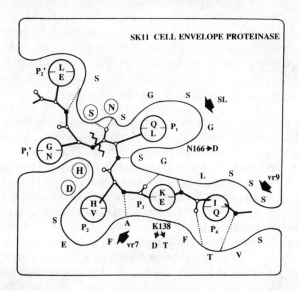

Fig. 5. Model of the substrate binding region in the catalytic domain of the SK11 proteinase containing two different α_{-s1} casein fragments. For further explanation see text.

4. ENGINEERING THE SK11 PROTEINASE

Following the cloning, expression and characterization of the SK11 *prt* genes and the emerging modelling studies a great variety of genetic and protein engineering experiments have been performed to determine the structure-function relation of this enzyme, which is the largest representative of the serine proteinases of the subtilase family[45]. The techniques used to achieve this vary from subcloning, overexpression and deletion experiments to site-specific and cassette mutagenesis, loop engineering and the production of hybrid enzymes. Here a summary will be given of the recent engineering approaches in our laboratory that have resulted in engineering SK11 proteinase activity, production, location, processing, size, specificity and thermostability.

Engineering the catalytic triad of PrtP

A crucial experiment to assess the applicability of the model of the SK11 proteinase was the engineering of the proposed active site residues (Fig. 4). This was addressed by generating mutant SK11 proteinase S433A that appeared to be devoid of caseinolytic activity[14]. In addition, this experiment identified S433 as the likely site for the inhibitors PMSF and DFP that covalently bind serine residues and are known to inhibit the lactococcal proteinases. The active site mutant S433A proved to be useful in elucidating the autocatalytic and intermolecular processing of the SK11 proteinase (see below)[14,48]. A similar active site mutant, D33A, has been created in the Wg2 proteinase yielding essentially the same results and confirming the proposed model[49].

Engineering the processing of the SK11 proteinase

In the course of the biochemical purification it appeared that the SK11 and other lactococcal proteinases are subject to N- and C-terminal proteolysis. Except for the cleavage of the signal peptide by the signal peptidase, the other degradation events are autoproteolytic and may activate, inactivate or not affect the proteolytic activity. Four of these autoproteolytic sites have been mapped at the both ends of the SK11 proteinase and appear to be subject to cleavage in a hierarchial order, as deduced from the intensity and activity of the resulting products.

Inactive forms of the SK11 proteinase obtained by site-directed mutagenesis of the active site (S344A) or in strains without a functional *prtM* gene, contained a N-terminal extension. The N-terminal sequence of these proteins could not be determined by Edman degradation in contrast to the wild-type proteinase that starts with the sequence at position 1 (Fig. 3)[14,33,42]. These results indicate that the primary autoproteolytic site is located at this position. In our current model of processing, cleavage at this site activates the proteinase by eliminating the N-terminal pro-region that probably acts as an intramolecular chaperon (Fig. 3)[15]. Recent experiments in which inactive, N-terminally extended proteinase was incubated with stoichiometric amounts of active SK11 proteinase showed the removal of the pro-peptide, indicating that autoproteolytic activation is probably an intermolecular event[48].

A well-known strategy for the purification of the lactococcal proteinases is release from the cell-envelope by incubation in a calcium-free buffer[50]. It has been shown that this is a result of intermolecular autodigestion and involves C-terminal processing[14,51]. In an indirect approach using C-terminally truncated proteinases obtained by 3' deletions in the *prtP* gene, this processing site has been located between residues 1127 and 1272 (site R for release, Fig. 3)[52].

Two processing sites that lead to inactivation of the SK11 proteinase have recently been mapped. A major N-terminal processing site that leads to an inactive proteinase, has been mapped in the surface loop at position 205-219 confirming its location (SL, Figs. 3-5)[46,48]. Protein engineering experiments aimed at deleting or substituting this major autoproteolysis site were unsuccessful since the surface loop contributed to activity and specificity (see below)[48]. C-terminal sequencing of an inactive degradation product suggested that another autoproteolysis site is located at bond 623-624 (site X, Fig. 3)[53].

Engineering production, size and location of the SK11 proteinase

By using a series of multicopy vectors the SK11 *prtP* and *prtM* genes have been cloned in *L. lactis*[35,39]. In addition, constructions have been made in which the expression of the *prtP* gene was lowered by substituting the promoter region[40]. The resulting plasmids were tested for the production of active proteinase and a maximal overproduction of approximately three-fold could be realized. Interestingly, the growth rate in milk appeared to be dependent on the level of proteinase production. This indicates that the caseinolytic activity of the cell-envelope located proteinase is the rate-limiting step for growth in milk and confirms earlier physiological experiments[54].

By deleting the C-terminal membrane anchor by generating a series of 3'-deletions in the *prtP* gene or by introducing a charged residue into the hydrophobic core of this stop-transfer sequence, it was possible to completely secrete the SK11 proteinase[40,42,55]. In the case of deletions this also resulted in C-terminally truncated proteinases with different size that retained proteolytic activity[52]. Since the secretion did not affect the growth rate in milk it was concluded that the location *per se* of the proteinase does not affect its action on casein[40]. An important spin-off from those studies was the possibility to obtain large amounts of the proteinase by concentrating the supernatant of *L. lactis* cells grown in cheap, whey permeate-based media[38].

Identification and engineering of substrate binding region 1

The proteinases of SK11 and Wg2 differ in only 44 out of 1902 residues[33]. Nevertheless, the caseinolytic specificity of these two proteinases differs considerably[27-29]. Various hybrid proteinases were generated, incorporating different segments of the two proteinases that all resulted in proteolytically active enzymes[55]. A major effect on the caseinolytic specificity was obtained by substituting the N-terminal 173 resdidues including the substrate binding region 1 from the catalytic domain of the Wg2 proteinase into the SK11 proteinase. This results in a proteinase with five mutations in the substrate binding region 1 (S131T, K138T, A142S, V144L and N166D) that shows a novel specificity towards casein[55]. This is illustrated in Fig. 6 showing the action of one of such hybrid proteinases (designated abCd) on the casein fragment α_{s1}-casein (1-23)[29].

Having established experimental evidence for the presence of a specificity region, designated region 1, that includes residues at both sides of the putative substrate binding cleft (Figs. 4 and 5), it was of interest to determine whether single amino acid substitutions would generate the expected specificity. It was predicted from the model that changing only the charge of the residues 138 and 166, located at either side of the binding cleft, would also affect the specificity of the SK11 proteinase (Fig. 5). This was experimentally confirmed by generating the mutant SK11 proteinases N166D and the double mutant K138D/A137G, the latter resulting in changing the specificity from a PIII-type to a PI-type (see Fig. 6).

Since autoproteolysis leads to inactivation of the proteinase, the specificity and stability are intimately coupled. Interestingly, the thermostability of the engineered proteinases could be altered and improved by several degrees in mutant proteinases such as the N166D and the K138D/A137G proteinase[47].

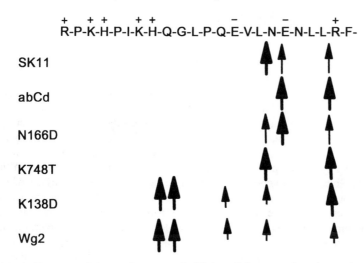

Fig. 6. Cleavage of the α_{s1}-fragment (1-23) by wild-type and engineered proteinases. The size of the arrows is related to the cleavage preference.

Identification and engineering of substrate binding region 2

The engineering experiments with the hybrid proteinases also allowed to define a second specificity region, designated region 2, in the C-terminal part of the proteinase (Fig. 3)[55]. This region has no equivalent in subtilases and therefore can not be modelled. It includes two residues that are positively charged in the SK11 but not in Wg2 (R747L and K748T) and contributes to the cleavage specificity towards casein, casein fragments or small model peptide substrates (Fig. 5)[29,55]. In subsequent engineering experiments it has been established that residue 748 contributes to the caseinolytic specificity since mutant K748T showed a new specificity (Fig. 6). It is hypothesized that region 2 folds back to the active site and contributes to the specificity by electrostatic interactions with the substrate. In addition, since the mutant K748T shows reduced stability, an indirect effect on the folding of the proteinase can not be excluded.

Identification and engineering of substrate binding region 3

A cassette mutagenesis approach was adopted in an attempt to engineer the stability of the SK11 proteinase by analyzing the function of the surface loop SL (position 205-219) that contains an autoproteolytic site (see above). A deletion mutant which lacks the 14 residues of SL was constructed and subsequently used to introduce various insertion cassettes coding for the original loop with three mutations (E205S/E218T/M219S; triple mutant) or for neutral spacers of 1,4, 7, or 16 serine residues. It appeared that the presence of residues 205-219 is essential for proteolytic activity since only the triple mutant retained caseinolytic activity. Unexpectedly, the triple mutant was found to be defective in C-terminal autoprocessing at site R (Fig. 3). Although an altered accessibility of site R can not be ruled out completely, this result suggests that SL forms a third region, designated region 3, that contributes to substrate specificity (Fig. 3).

Engineering of loop 238-338 of the SK11 proteinase

The residues 238-388 in the SK11 proteinase constitute the largest insert in an external loop, and it is sufficiently large to constitute an additional internal domain (see Figs. 3 and 4). A similar but less conserved domain is found in the *S.pyogenes* endopeptidase and the *B.subtilis* minor protease. A mutant that lacked residues 238-288 was constructed and produced in *L.lactis*[56]. This mutant strain showed a two-fold reduced growth rate in milk indicating that the large domain is not essential for activity and its removal does not inhibit folding. By using monoclonal antibodies that had been raised against the lactococcal proteinase we were able to show that the antigenic determinant mAB-I, previously mapped in the segment 215-604, is at least partially located in the loop 238-388[56,57]. Since this epitope is most likely to be located on the surface of the catalytic domain, it represents a good candidate for inserting heterologous antigenic determinants. Presently, we are studying the possibility of using *L.lactis* strains expressing such mutated SK11 proteinases as carriers for foreign antigenic determinants in new vaccination programmes.

4 ENGINEERING OF AMINOPEPTIDASE N

The general aminopeptidase N has been purified to homogeneity allowing the generation of antibodies and determination of the N-terminal sequence[23,58,59]. It is a monomer with a size of approximately 95 kD and shows a broad substrate specificity, degrading several di-, tri- and larger peptides by hydrolysis of the N-terminal amino acid but has no endopeptidase or carboxypeptidase activity. Specifically, lysyl- and leucyl-para-nitroanilides are cleaved and these synthetic substrates as well as their ß-naphthylamide derivatives have been used as chromogenic substrates during screening, purification, and overproduction of PepN.

Structure of the *pepN* gene and its product

The structural *pepN* gene has been identified in a lambda library of *L.lactis* MG1363 DNA using antibodies prepared against the purified PepN[59]. The nucleotide sequence of the *pepN* gene has been determined and its transcription has been analyzed[60]. A similar sized gene has been analyzed from *L.lactis* Wg2 that differs in only a few nucleotides[61]. The *pepN* gene shows a monocistronic organization and is preceded by a consensus promoter sequence (Fig. 7).

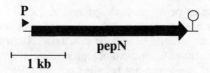

Fig. 7. Sequence organization of the *pepN* gene.

The deduced primary sequence of the 846-residue PepN is homologous to Zn-metalloenzymes from various pro- and eukaryotes, with the highest similarity to the mammalian aminopeptidase N[60]. A highly conserved segment comprising residues 281-321 shows homology with the thermolysin family of Zn-dependent neutral proteinases for which various 3-D structures are known. This homology analysis suggests that residues H288, H292 and E311 are the Zn-ligands whilst E289 is involved in catalysis. The N-terminal sequence determined from the purified PepN corresponds to that deduced from the gene sequence and indicates that the protein does not contain a signal peptide[59]. In addition, no transmembrane sequences were found in the deduced primary structure which is compatible with its intracellular location as revealed by fractionation and immunogold labelling studies[23,58,59].

PepN has debittering activity

Although caseins form a nutritionally well-defined and valuable protein source, their use can be extended by predigestion with proteolytic enzymes. However, in many cases specific hydrophobic peptides are generated that cause bitter off-flavors. Because of its broad substrate specificity, aminopeptidase N has been analyzed for the capacity to debitter a tryptic hydrolysate of ß-casein[24]. The oligopeptides in such an hydrolysate before and after incubation with aminopeptidase N were isolated and subsequently identified by N- and C-terminal amino acid sequencing, amino acid composition analysis, and on line liquid chromatography - mass spectrometry. In this way it could be established that several tryptic fragments, including those originating from the bitter C-terminal end of the ß-casein, were completely hydrolysed by aminopeptidase N[24]. In line with this enzymatic activity is the observation that the bitter score of the tryptic hydrolysate after treatment with aminopeptidase N was considerably less than before treatment. Although this demonstrates the debittering potential of aminopeptidase N, it is evident from the amounts of enzyme needed to obtain debittering that commercially viable processes require a cheap source of the enzyme that only can be obtained by genetic engineering.

Engineering of PepN

In the first attempt to engineer the *pepN* gene it has been cloned in *E.coli* under control of several inducible promoters[59]. This resulted in the significant overproduction of aminopeptidase N that could be purified from this heterologous host and used to raise antibodies. These antibodies did not show the cross-reactivity to other lactococcal proteins that were found with antibodies against aminopeptidase N purified from *L.lactis* and hence have been used in various localization experiments[23].

Subsequently, the *pepN* gene has been cloned in *L.lactis* using a high copy number vector. This resulted in an overproduction of approximately 25-fold in the homologous host. Since the generated strain contains a transferable antibiotic-resistance gene located on the *pepN*-expressing plasmid, the recently developed NIZO food-grade marker system based on complementation of the *lacF* gene was incorporated[2,4,62]. The resulting plasmid expressing the *pepN* gene has been obtained by self cloning and contained exclusively DNA originating from *L.lactis*. A *L.lactis* strain harboring this plasmid has now been excluded from the existing biosafety legislation and has potential to be developed for the food-grade production of aminopeptidase N.

By replacement recombination a mutant of *L. lactis* MG1363 has been constructed that contains a deletion in the chromosomal *pepN* gene[63]. This strain did not produce aminopeptidase N activity and was used to demonstrate that there are other broad host-range aminopeptidases in *L. lactis*[23]. This may explain the observation that this PepN-deficient strain showed almost the same growth and acidification rate in milk as the wild-type strain[63]. The thus constructed PepN-deficient strain could be used as a suitable host in experiments aimed at overexpressing mutated *pepN* genes in order to further study the structure-function relationship of aminopeptidase N.

5. CONCLUDING REMARKS

In recent years the lactococcal cascade of proteolytic processes has developed into one of the best characterized systems involved in bacterial nitrogen metabolism. In this paper we have reviewed the engineering aspects of two important enzymes in the proteolytic cascade, the cell-envelope located proteinase and the general aminopeptidase N, that appear to be widespread in proteolytic systems of other lactic acid bacteria as has been recently reviewed[15]. Apart from deepening our insight in the structure-function relations of complex enzymes, these engineering studies have resulted in or have laid the basis for development of lactococcal starter strains with improved properties. It is expected that by using these strains in industrial processes not only novel or improved products may be developed, but also our understanding will be increased of the contribution of the proteolytic system to the final flavor and consistency of fermented dairy products. In this respect it is relevant to mention recent engineering studies that involve the substrate casein. Using a protein engineering approach we have been able to generate and overproduce mutant forms of ß-casein that have novel properties when incubated with a proteolytic enzyme, such as chymosin[64]. Although the development of transgenic cows producing mutant caseins will take a long time, it illustrates an alternative avenue that can be taken to improve existing and generate novel fermented dairy products.

ACKNOWLEDGEMENTS

Part of this work was supported by EC contract BIOT-CT91-0263 in the framework of the BRIDGE T-project 'Biotechnology of Lactic Acid Bacteria'.

REFERENCES

1. W.M. de Vos, J.J. Huis in 't Veld, and B. Poolman, Eds., Special Issue FEMS Microbiol. Rev., 1993, 12, 1.
2. W.M. de Vos and G. Simons, 'Genetics and Biotechnology of Lactic Acid Bacteria', M.J. Gasson and W.M. de Vos, Eds., Chapman and Hall, London, 1994, p.52.
3. W.M. de Vos, in 'Harnessing Biotechnology in the 21st Century', R. Lodisch & A. Bose, Eds., American Chemical Society, Washington, 1992, p.524.
4. W.M. de Vos, I. Boerrigter, R.J. van Rooijen, B. Reiche and W. Hengstenberg, J. Biol. Chem., 1990, 265, 22554.
5. R.J. van Rooijen, S. van Schalkwijk and W.M. de Vos, J. Biol. Chem., 1991, 266, 7176.
6. J. van Rooijen and W.M. de Vos, J. Biol. Chem., 1991, 265, 18499.
7. R.J. van Rooijen and W.M. de Vos, Protein Engng., 1993, 6, 210.
8. W.M. de Vos and E. Vaughan, FEMS Microbiol. Rev., 1994, 13, in press.
9. O.P. Kuipers, H.S. Rollema, W.M.G.J. Yap, H.J. Boot, R.J. Siezen and W.M. de Vos, J. Biol. Chem. 1992, 267, 2430.
10. H.M. Dodd, N. Horn, Z. Hao and M.J. Gasson, Appl. Environ. Microbiol. 1992, 58, 3683.
11. J.R. van der Meer, H.S. Rollema, O.P.Kuipers, R.J. Siezen and W.M. de Vos, J. Biol. Chem. 1994, 3555.
12. P.J.G. Rauch, O.P.Kuipers, R.J. Siezen and W.M. de Vos 'Bacteriocins of Lactic Acid Bacteria', L. de Vuyst and J. Vandamme, Eds., Chapman and Hall, London, 1994, p. 223.
13. J. Kok, FEMS Microbiol. Rev., 1990, 87, 15.
14. W.M. de Vos, I. Boerrigter, P. Vos, P. Bruinenberg and R.J. Siezen, 'Genetics and Molecular Biology of Streptococci, Lactococci and Enterococci', G.M. Dunny, P.P.Cleary and L.L. McKay, Eds., American Society for Microbiology, Washington, DC., 1991, p.115.
15. J. Kok and W.M. de Vos, 'Genetics and Biotechnology of Lactic Acid Bacteria', M.J. Gasson and W.M. de Vos, Eds., Chapman and Hall, London, 1994, p.169.
16. J.E. Citti, W.E. Sandine and P.R. Elliker, J. Dairy Sci., 1965, 48, 1253.
17. J.R. Van der Meer, J. Polman, M.M. Beerthuyzen, R.J. Siezen, O.P. Kuipers and W.M. de Vos, J. Bacteriol., 1993, 175, 2578.
18. R.J. Siezen, 'Subtilisin Enzymes', Ch. Betzel and R. Bott, Eds. Plenum Press, New York, 1994, in press.
19. S. Tynkkynen, G. Buist, E. Kunji, J. Kok, B. Poolman and G. Venema, J. Bacteriol., 1993
20. E. Smid, R. Plapp and W.N. Konings, J. Bacteriol. 1989, 171, 6135
21. W.N. Konings, B. Poolman and A.J.M. Driessen (1988) CRC Critical Reviews in Microbiology, 1989, 16, 419.
22. J. Stadhouders,, L. Toepoel and J.T.M. Wouters, Neth. Milk Dairy J., 1988, 42, 182.

23. R. Baankreis, 'The role of lactococcal peptidases in cheese ripening', Academic Thesis, University of Amsterdam, 1992, Amsterdam, The Netherlands.
24. P.S.T. Tan, T.A.J.M. van Kessel, F.L.M. van de Veerdonk, P.F. Zuurendonk, A.P. Bruins and W.N. Konings, Appl. Environ. Microbiol., 1993, 59, 1430.
25. F.A. Exterkate, Neth. Milk Dairy J., 1976, 30 95-105.
26. T.D. Thomas and G.G. Pritchard, FEMS Microbiol. Rev., 1987, 46, 245.
27. S. Visser, F.A. Exterkate, K.J. Slangen, and G.J.C.M. de Veer, Appl. Environ. Microbiol., 1986, 52, 1162.
28. S. Visser, J. Dairy Sci, 1993, 76, 329.
29. F.A. Exterkate, A.C. Alting and P. Bruinenberg, Appl. Environ. Microbiol., 1993, 59, 3640.
30. G.G. Pritchard and T. Coolbear, FEMS Microbiol. Rev., 1993, 12, 179.
31. J.R. Reid, T. Coolbear, C.J. Pillidge and G.G. Pritchard, Appl. Environ. Microbiol., 1994, 60, 801-806.
32. J. Kok, K.J. Leenhouts, A.J. Haandrikman, A.M. Ledeboer, and G. Venema, Appl. Environ. Microbiol., 1988, 54, 231.
33. P. Vos, G. Simons, R.J. Siezen & W.M. de Vos, J. Biol. Chem., 1989, 264, 13579.
34. S. Visser, G. Hup., F.A. Exterkate and J. Stadhouders, Neth. Milk Dairy J., 1983, 37, 169.
35. W.M. de Vos, FEMS Microbiol. Rev., 1989, 46, 281.
36. W.M. de Vos, H.U. Underwood and F.L. Davies, FEMS Microbiol. Lett., 1984,
37. W.M. de Vos and F.L. Davies, 'Third European Congress on Biotechnology' Verlag Chemie, Basel, 1984, Vol. III, p. 202.
38. W.M. de Vos, P. Vos, H. de Haard and I. Boerrigter, Gene, 1989, 85, 169.
39. W.M. de Vos, P. Vos, G. Simons and S. David, J. Dairy Sci., 1989, 72, 3398.
40. P. Bruinenberg, P. Vos and W.M. de Vos, Appl. Environ. Microbiol., 1992, 58, 78.
41. J. Marugg, P. Bruinenberg, P. Laverman, R. van Kranenburg and W.M. de Vos, 1994, manuscript in preparation.
42. P. Vos, M. van Asseldonk, F. van Jeveren, R.J. Siezen, G. Simons and W.M. de Vos, J. Bacteriol., 1989, 171, 2795.
43. A.J. Haandrikman, J. Kok and G. Venema, J. Bacteriol., 1993, 173, 4517.
44. A.J. Haandrikman, C. van Leeuwen, J. Kok, P. Vos, W.M. de Vos and G. Venema, Appl. Environ. Microbiol., 1990, 56, 1890.
45. R.J. Siezen, W.M. de Vos, J.A.M. Leunissen, and B. Dijkstra, Prot. Engng, 1991, 4, 501
46. P.G. Bruinenberg, P. Vos, F.A. Exterkate, A.C. Alting, W.M. de Vos and R.J. Siezen 'Stability and stabilization of enzymes', W.J.J. van den Tweel, A. Harder and R.M. Buitelaar, Eds., Elsevier Science Publishers, Amsterdam, 1993, p. 231.

47. R.J. Siezen, P.G. Bruinenberg, P. Vos, I. van Alen-Boerrigter, M. Nijhuis, A.C. Alting, F.A.Exterkate and W.M. de Vos, Prot. Engng., 1993, 6, 927.
48. P. Bruinenberg, W.M.de Vos and R.J. Siezen, Biochem. J., 1994, in press.
49. A.J. Haandrikman, R. Meesters, H. Laan, W.N. Konings, J. Kok and G. Venema, Appl. Environ. Microbiol., 1991, 57, 1899.
50. O.E. Mills and T.D. Thomas N. Z. J. Dairy Sci. Technol., 1978, 13, 209.
51. H. Laan and W.N. Konings, Appl. Environ. Microbiol., 1989, 55, 3101.
52. P. Bruinenberg, M. Nijhuis, P. Vos, P. Laverman, H. Laan, R.J. Siezen and W.M. de Vos, 1994, manuscript in preparation.
53. P. Bruinenberg, W.M. de Vos and R.J. Siezen, unpublished results.
54. J. Hugenholtz, M. Dijkstra and H. Veldkamp, FEMS Microbiol. Ecol., 1987, 53, 149.
55. W.M. de Vos, I.J. Boerrigter and P.Vos, unpublished results.
55. P. Vos, I.J. Boerrigter, G. Buist, A.J. Haandrikman, M. Nijhuis, M.B. de Reuver, R.J. Siezen, G. Venema, W.M. de Vos, and J. Kok, 1991, Prot. Engng. 4, 479.
56. P. Bruinenberg, P. Doesburg, A.C. Alting, F.A. Exterkate, W.M. de Vos and R.J. Siezen, 1994, submitted for publication.
57. H. Laan, J. Kok, A. Haandrikman, G. Venema and W.N. Konings, Eur. J. Biochem., 1992, 204, 815.
58. P.S.T. Tan and W.N. Konings, Appl. Environ. Microbiol., 1990, 56, 526.
59. I.J. van Alen-Boerrigter, R. Baankreis and W.M. de Vos, Appl. Environ. Microbiol., 1991, 57, 2555.
60. P.S.T. Tan, I.J. van Alen-Boerrigter, B. Poolman, R.J. Siezen, W.M. de Vos and W.N. Konings, FEBS Lett. 1992, 306, 9.
61. P. Strøman, Gene, 1992, 113, 107.
62. W.M. de Vos, Eur. Patent Appl., 1989, 0 355 036
63. I.J. van Alen-Boerrigter and W.M. de Vos, unpublished results.
64. G. Simons, W. van den Heuvel, T. Reynen, A.Frijters, G. Rutten, C. Slangen, M. Groenen, W.M. de Vos and R.J. Siezen, Prot. Engng., 1993, 6, 763.

Protein Engineering and Preliminary X-Ray Analysis of CHY155-165RHI Loop Exchange Mutant

J. E. Pitts[1], P. Orprayoon[1], P. Nugent[1], R. V. Dhanaraj[1], J. B. Cooper[1], T. L. Blundell[1], J. Uusitalo[2], and M. Penttilä[2]

[1] DEPARTMENT OF CRYSTALLOGRAPHY, BIRKBECK COLLEGE, MALET STREET, LONDON WC1E 7HX, UK

[2] THE BIOTECHNICAL LABORATORY, VTT, SF-02151, ESPOO, FINLAND

1 INTRODUCTION

Knowledge of the three-dimensional structures of proteins and molecular modelling techniques provides a powerful approach for the design of novel biomolecules with specifically engineered properties. Structure-based design is a central feature of protein engineering, which can be described as a multidisciplinary design cycle[1] (Figure 1). Recombinant DNA and biochemical approaches are used to prepare proteins and then the combined structural techniques of X-ray analysis, NMR and computational modelling can be utilised to determine and model three-dimensional structures. In one cycle mutant proteins with altered functions, specificities or stabilities can be designed and expressed.

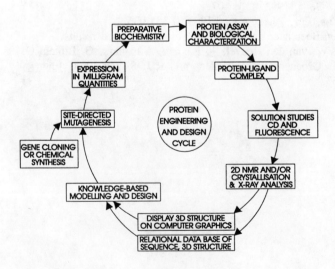

Figure 1 Protein engineering design cycle

Usually the determination of the three-dimensional structure of the target protein is a rate-limiting step in this process. However, quite often another protein with a common fold can be identified and much can be learnt from comparative modelling using the sequence of the target protein and the three-dimensional structures of one or more related proteins[2]. Even when the structure of the target protein has been determined, the comparative study of families of homologues and their ligand complexes can be very useful in understanding specificity and stability, and in designing novel proteins and ligands. The research carried out at Birkbeck on aspartic proteinases provides a good example to illustrate these themes.

Aspartic proteinases form a family of endopeptidases belonging to a wide range of biological species with varying substrate specificities. The members of the family are of considerable commercial importance since enzymes such as chymosin and *Mucor pusillus* pepsin have been exploited by the food industry in cheese and soya processing, while renins, cathepsins and the retroviral proteinases are prime targets of the pharmaceutical industries in drug design. Several aspartic proteinases have been the subject of extensive X-ray structural analyses of both the native and inhibitor-bound forms. These high resolution structures provide an excellent database for the development of biomolecular design principles (Figure 2). The structures of about 40 peptidomimetic inhibitor complexes have provided a detailed definition of the shape and geometry of the specificity pockets; this knowledge is indispensable in the design of potent inhibitors of pharmacological interest.

This article describes progress around the protein engineering cycle using chymosin as an example. We describe the expression and structural analysis of one chymosin mutant; this is part of a programme designed to investigate the role of loop regions in the structure and stability of aspartic proteinases. The major theme is to produce chimeric proteins by replacing variable regions in one enzyme with the equivalent region, often of a different length and quite different sequence. These studies are truly multidisciplinary, but rely on the existence of many high resolution three-dimensional structures of various homologues.

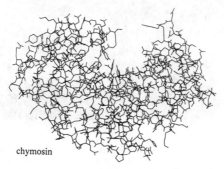

Figure 2 Alpha-carbon tracing of bovine chymosin

2 PROTEIN ENGINEERING

The protein engineering programme at Birkbeck seeks to develop generic methods based on a design cycle involving biochemical preparation and characterization, determination and comparative analysis of three-dimensional structures, rule-based design, site-directed mutagenesis and expression of the mutants. The research has concentrated mainly on chymosin. The ability of chymosin to cleave specifically the Phe^{105} - Met^{106} bond in κ-casein to initiate milk clotting makes it useful for manufacturing cheese. The objective of the protein engineering exercise is to target alterations to a number of properties including size, specificity and stability. For instance, in the case of alteration of size, the aim is to reduce the length of surface loops to allow more efficient penetration of complex substrates used in milk processing. By comparing the three-dimensional structure of chymosin B[3] with those of the other members of the aspartic proteinase family, a surface loop ($q^N r^N$) formed by residues 155-165 (pepsin numbering) was initially targeted for modification by replacement with homologous sequences from the other members of the family (Figure 3) as well as with other modelled loops of variable chain length. The sections below describe the expression, crystallization and structural analysis of such a loop exchange mutant where the target loop was substituted by the corresponding loop from *Rhizopus chinensis* pepsin[4,5].

<u>Table 1</u> Comparison of the chymosin 155-165 loop with a selection of aspartic proteinases analyzed by X-ray analysis

CHYMOSIN	M	D	R	N	–	–	–	–	G	Q	E	S	M	L
RHIZOPUSPEPSIN	I	G	K	A	K	N	–	–	G	G	G	G	E	L
ENDOTHIAPEPSIN	M	G	Y	H	–	–	–	–	–	A	P	G	T	L
MUCORPEPSIN	M	N	T	N	–	–	–	–	–	D	G	G	Q	L

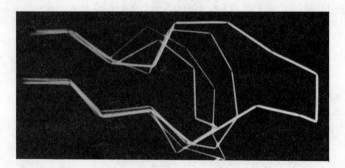

<u>Figure 3</u> A series of surface loops (155-165) from aspartic proteinases are overlaid to indicate potential structural differences

Site-directed mutagenesis of chymosin

The mutation Chy155-165Rhi was introduced into the chymosin cDNA using a phosphorothioate based site-directed mutagenesis method[6]. The mutagenic 61-mer oligonucleotide (Table 2) was synthesised using phosphoramidite chemistry and purified by reverse phase chromatography on an FPLC system. The Eco RI-Msc I restriction fragment containing the mutation was cloned into the vector pAMH104E-ΔBam HI in the correct orientation. The DNA was then cleaved with Eco RI, dephosphorylated and combined with the large Eco RI fragment of pAMH104[7] (Figure 4) to produce the mutant expression vector pCHY-RHI115. The correct insert was confirmed by DNA sequencing using the M13 dideoxy chain termination method of Sanger[8] (Figure 5).

Table 2 Sequence comparison of chymosin and Rhizopuspepsin 155-165 loops and the changes introduced by the mutagenic 61-mer oligonucleotide, base changes are indicated in bold

CHYMOSIN
155
MET ASP ARG ASN GLY GLN --- --- GLU SER MET LEU

5' ATG GAC AGG AAT GGC CAG   GAG AGC ATG CTC 3'

5' AT**C** GGC **AAG** GCT **AAG AAC GGA GGT** GGC GGA G**A**G CTC 3'

ILE GLY LYS ALA LYS ASN GLY GLY GLY GLY GLU LEU
155
RHIZOPUSPEPSIN

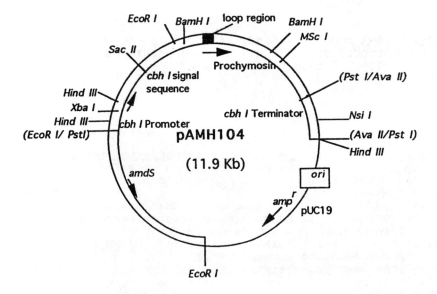

Figure 4 Expression vector pAMH104 for chymosin production in *Trichoderma reesei*

3' GACAAGAGCCAAATGTA
              GCCGTTCCGATTCTTGCCTCCACCGCCTCT
                                           CGAGTGCACCCCC 5'

Mutagenic 61-mer used to introduce the Chy155Rhi

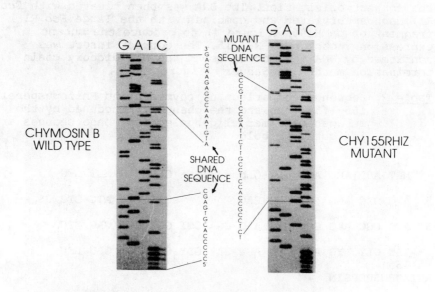

Figure 5   DNA sequencing indicating the successful introduction of the Rhizopuspepsin loop 155-165 into chymosin

Expression of chymosin mutants in Trichoderma reesei

The filamentous fungus *Trichoderma reesei* (strain Rut C30) was transformed with the expression plasmid pCHY-RHI155 containing the mutant DNA and plated onto selective minimal agar containing acetamide as sole nitrogen source (Fig. 6). The correctly folded zymogen undergoes secretion into the culture medium under the direction of the strong cellobiohydrolase I (*cbh*I) promoter. The low pH of the growth medium leads to autocatalytic activation of the enzyme. The best producer/transformant was grown at the 10l scale in an Applikon bioreactor using cellulose as the sole carbon source. Chymosin activity was detectable in the liquid culture after three days of growth. The milk clotting activity was maximal at day five, while the amount of total secreted proteins continued to increase. The yield of the mutant Chy155-165Rhi was 27 mg/l.

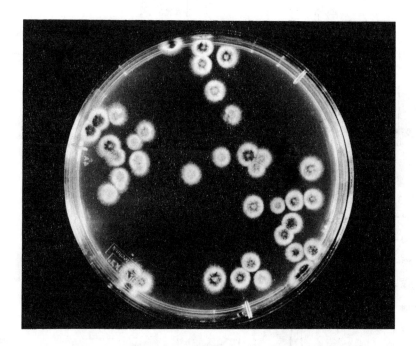

Figure 6 Selection by acetamide. Transformants with CHY155-165Rhi (normal T. reesei gave completely clear plates showing no growth).

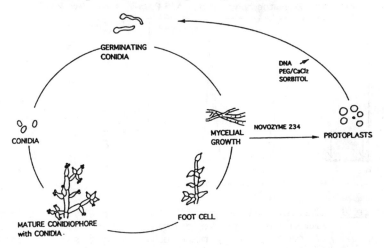

Figure 7 Life cycle of *Trichoderma reesei*

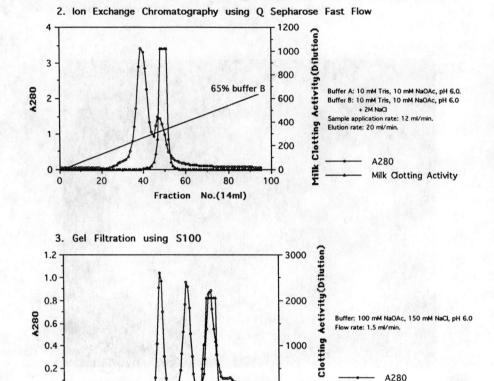
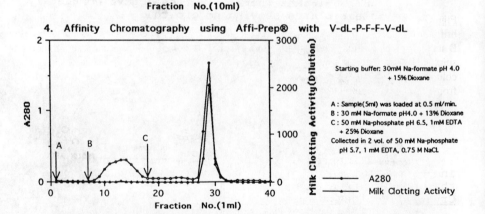

Figure 8 The four step purification for Chy155—165Rhi; ammonium sulphate precipitation, anion-exchange chromatography, molecular weight sieving by gel filtration and finally affinity-binding on a **V-dL-P-F-F-V-dL** inhibitor column.

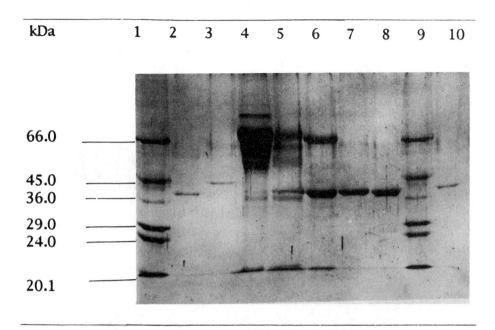

Figure 9 11% SDS-PAGE of the chymosin mutant Chy155-165Rhi stained with Coomassie brilliant blue. Lanes 1 and 9: low molecular weight markers; lanes 2 and 10: calf chymosin B standard; lane 3: trichodermapepsin; lane 4: spent medium; lane 5: sample after ammonium sulphate fractionation and dialysis; lane 6: pooled active material from Q Sepharose Fast Flow column; lane 7 pooled active material from the S100 column; lane 8: pooled active material from affinity chromatography.

Enzyme kinetics using a synthetic chromogenic peptide indicated that the activity of the mutant is similar to that of wild-type chymosin B.

Crystallization and data collection

The purified mutant was concentrated to 10 mg/ml in sodium phosphate buffer at pH 5.6 and crystallised by the hanging drop vapour diffusion method using sodium chloride as the precipitant. Plate-like crystals of dimensions 0.20 X 0.15 X 0.05 mm, grown over a period of 3 to 4 weeks, were found to be isomorphous with the native crystals (Figure 10). X-ray intensity data were collected by the oscillation technique with a MAR RESEARCH image plate detector using the Daresbury Laboratory Synchrotron Radiation Source. Although the X-ray diffraction data extended to 2.2 Å resolution, the images showed splitting due to the existence of multiple crystals that were impossible to separate. The intensities were weak due to the small size of the crystals at resolutions beyond 2.5 Å. Oscillation images from two differently-oriented crystals were processed using the CCP4 software suite and merged to a unique set of reflections (R_{merge} 12.4%) extending to a resolution of 2.5 Å.

<u>Figure 10</u> Crystals of the Chy155-165Rhi mutant which are isomorphous to native chymosin

<u>Refinement and results</u>

The atomic coordinates of native chymosin B[4], after the removal of the mutated loop, were used to obtain the initial difference Fourier maps. Interactive model building on an Evans & Sutherland PS300 with the graphics software FRODO[9], followed by least squares refinement cycles using RESTRAIN, gradually revealed the atomic positions of the residues in the altered loop. The current crystallographic agreement factor and correlation coefficient for the refined model are 0.206 and 0.917 respectively, for reflections in the resolution range 8.0-2.5 Å. The superposition of the equivalent loop from rhizopuspepsin on that of the mutant is shown for comparison (Figure 11). Although the overall conformation of this mutant enzyme is similar to that of chymosin, clear conformational differences are found for the swapped loop when compared with the same sequence in the rhizopuspepsin structure. As the loop sequence is identical in the rhizopuspepsin and Chy155-165Rhi mutant, this indicates a strong influence of the local side-chain environment on the conformation, most significantly at residues Gly 161 and Gly 162. Interestingly the mutant loop exhibits the presence of a water molecule in the vicinity of the hairpin (Figure 11). This water molecule stabilises the β-hairpin conformation by making hydrogen bonds with the main-chain nitrogen and carbonyl oxygen atoms of the residues in the loop. This is in contrast to the native rhizopuspepsin which does not have a water molecule in the equivalent position.

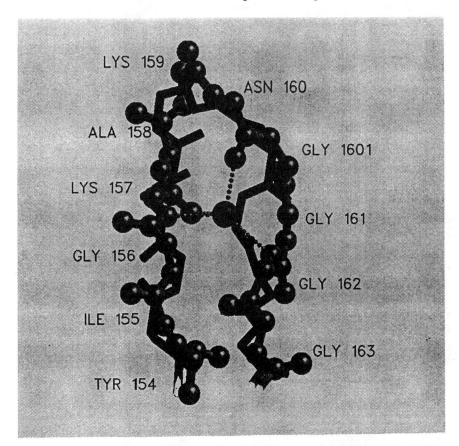

Figure 11 The main-chain atoms corresponding to the mutated loop (in ball and stick representation) and the water molecule (W) which stabilises the β-hairpin conformation of the mutant; the equivalent loop in rhizopuspepsin structure (smooth cylindrical joints) has been superposed for comparison. The pepsin numbering scheme has been used for labelling.

3 CONCLUSION

The loop mutant Chy155-165Rhi was produced in the filamentous fungus *Trichoderma reesei* and exported into the growth medium to yield a correctly folded and highly active product. Structural and kinetic analyses confirm that the mutant adopts a native fold. Therefore surface loops may be successfully swapped between enzymes of different species without loss of catalytic activity, although they might be subtly altered by the local environment provided by the recipient protein. Such studies will provide further insights into the evolution of loops and their impact on structure and function.

References

1. J.E. Pitts, V. Dhanaraj, C.G. Dealwis, D. Mantafounis, P.G. Nugent, P. Orprayoon, J.B. Cooper, M. Newman and T.L. Blundell, Scan. J. of Clinical and Laboratory Investigation, 1992, 52, 39.

2. A. Sali, J.P. Overington, M.S. Johnson and T.L. Blundell, TIBS, 1990, 235.

3. M. Newman, M. Safro, C. Frazao, G. Kahn, A. Zolanov, I.J. Tickle, T.L. Blundell and N. Andreeva, J. Mol. Biol., 1991, 221, 1295.

4. K. Suguna, R.R. Bott, E.A. Padlan, E. Subramanian, S. Sheriff, G.H. Cohen and D.R. Davies, J. Mol. Biol., 1987, 196, 877.

5. V. Dhanaraj, J.E. Pitts, P.G. Nugent, P. Orprayoon, J.B. Cooper, T.L. Blundell, J.M. Uusitalo and M.E. Penttilä, 5th International conference on Aspartic Proteinases, 1993, Abstract P47.

6. J.W. Taylor, J. Ott and F. Eckstein, Nucleic Acids Res., 1985, 13, 8764.

7. J.E. Pitts, J.M. Uusitalo, D. Mantafounis, P.G. Nugent, D. Quinn, P. Orprayoon and M Penttilä, J. Biotechnology, 1993, 28, 69.

8. F. Sanger, S. Nicklen and A.R. Coulson, Proc. Natl. Acad. Sci. USA, 1977, 74, 5463.

9. A.T. Jones, J. Appl. Cryst., 1978, 11, 268.

Peptidases from Lactococci and Secondary Proteolysis of Milk Proteins

Francis Mulholland

BIOTECHNOLOGY AND ENZYMOLOGY DEPARTMENT, INSTITUTE OF FOOD RESEARCH, READING LABORATORY, EARLEY GATE, WHITEKNIGHTS ROAD, READING RG6 2EF, UK

1. INTRODUCTION

Lactic acid bacteria, such as lactococci, play an essential role in the manufacture of cultured dairy products such as cheese. They are primarily responsible for the acidification of the milk, the key stage in the manufacture of such products, and are also considered to be directly involved in the formation of the characteristic flavour notes in cheeses during ripening. The economic importance of cultured dairy products, and manufacturers demands for starter cultures that will produce a consistent product in the modern high throughput manufacturing facility has led to an in-depth investigation of these organisms, their physiology, biochemistry and genetics.

The lactococci used for cheese manufacture are nutritionally fastidious organisms, requiring, amongst other nutrients, exogenous sources of amino acids in order to achieve the rate of growth in milk that is necessary for the acidification required in cheese manufacture. This amino nitrogen deficiency can be met by two routes:

1. Directly, by uptaking amino acids or small peptides present in the extracellular media into the cell.

2. Indirectly, if there are insufficient small nutrients, by hydrolysing larger proteins in the media to amino acids and peptides of a transportable size.

To achieve this, lactococci have developed a proteolytic system, consisting of a mixture of enzymes, both proteolytic and peptidolytic, and several amino acid and peptide transport systems, that is capable of providing these essential nutrients to the cell.

Milk as a growth medium has a limited supply of free amino acids. Kolstad and Law (1) reviewed data that milk cannot provide all the free amino acids such as Histidine, Leucine, Glutamate and Methionine, required for growth, and lactococci rely on their proteolytic enzymes to hydrolyse the exogenous milk proteins to supply these essential nutrients. Milk has a number of proteins (total protein 3.0-

3.5% [wt/wt]), of which the caseins represent about 80%. The four different types of casein found in milk, α_{s1}, α_{s2}, β and κ, and are organised into micelles to form soluble complexes. It has generally been accepted that the caseins are the major source of amino acids utilised by lactococci for growth in milk.

As well as their importance in growth, the proteolytic systems of starter cultures have also been implicated in the maturation processes in cheese, where it is thought that the lactococcal proteolytic and peptidolytic enzymes play an essential role in the formation of amino acids and small peptides important in cheese flavour, either directly or as the precursors of flavour notes.

2. PRIMARY HYDROLYSIS OF MILK PROTEINS

The first step in proteolysis of milk proteins by lactococci is the action of the cell wall associated proteinase on whole caseins. This is an extracellular event and the proteinase has been well characterised, both genetically and biochemically in the past decade and the subject of several reviews (2,3,4). The essential role that this enzyme plays was demonstrated by the restricted growth shown by proteinase negative strains, lacking the plasmid encoding the proteinase compared to proteinase positive strains when grown on milk. This limitation in growth was removed when the media was supplemented with the essential free amino acids or by introducing the proteinase gene into the negative mutant (5,6).

It has been suggested that since proteolysis does play an essential role in the rapid growth of lactococci, then the early forming casein peptides resulting from the action of the proteinase would probably be the source of the amino acid nutrients required by the lactococci after the initial use of the available free amino acids and small peptides (3). Table 1 shows the early forming casein-derived peptides produced by the proteinase in *in vitro* studies. These peptides do contain the essential amino acids required by lactococci, although it is difficult to determine in what quantity they are made available. It was demonstrated that optimal growth of Lactococcus lactis subsp, cremoris HP required the presence of both β-and κ-casein in the media (7) and it was further speculated that κ-casein hydrolysis may be a source of the essential amino acid histidine (8). It is perhaps significant that one of the early formed peptides, κ-casein 96-106, is histidine rich.

Table 1   Early forming peptides produced by the lactococcal proteinase on casein

Casein Peptide	Amino Acid Sequence
β-casein176-182[a]	Lys-Ala-Val-Pro-Tyr-Pro-Gln
β-casein183-193[a]	Arg-Asp-Met-Pro-Ile-Gln-Ala-Phe-Leu-Leu-Tyr
β-casein194-207[a]	Gln-Gln-Pro-Val-Leu-Gly-Pro-Val-Arg-Gly-Pro-Phe-Pro-Ile
β-casein194-209[a]	Gln-Gln-Pro-Val-Leu-Gly-Pro-Val-Arg-Gly-Pro-Phe-Pro-Ile-Ile-Val
κ-casein 96-106[b]	Ala-Arg-His-Pro-His-Pro-His-Leu-Ser-Phe-Met
κ-casein 161-169[b]	Thr-Val-Gln-Val-Thr-Ser-Thr-Ala-Val
α_{s1}-casein 143-148[c]	Ala-Tyr-Phe-Tyr-Pro-Glu
α_{s1}-casein 162-169[c]	Gly-Ala-Trp-Tyr-Tyr-Val-Pro-Leu
α_{s1}-casein 170-199[c]	Gly-Thr-Gln-Tyr-Thr-Asp-Ala-Pro-Ser-Phe-Ser-Asp-Ile-Pro-Asn-Pro-Ile-Gly-Ser-Glu-Asn-Ser-Glu-Lys-Thr-Thr-Met-Pro-Leu-Trp

References [a] (9)   [b](3)   [c](10)

3.   SECONDARY HYDROLYSIS

Whilst the primary hydrolysis shows the probable peptides that will go on to supply lactococci with essential amino nitrogen nutrients, an examination of secondary proteolysis processes is required to determine how this is likely to be achieved.

It was speculated in 1991 that the cell wall associated proteinase and the peptidases known at that time to be associated with lactococci possessed the specificities required to hydrolyse nearly all the potential peptide bonds found in β-casein (8). Whilst total hydrolysis of the caseins is not a requisite for growth of lactococci in milk, and clearly not occurring in cheese, studies on the secondary hydrolysis of casein derived peptides are now becoming to be the focus of attention. A complementary range of physiological, biochemical and genetic approaches are being employed to examine the roles played by these different peptidases both on growth and on ripening processes in cheese.

4. LACTOCOCCAL PEPTIDASES

Over the past decade the lactococcal proteinase has been extensively studied. In contrast, the study of the peptidases found in starter cultures, and their roles, both in growth of the organism and in food manufacture, is just beginning. Recently, however, a substantial advancement in our knowledge of the peptidases has occurred. Table 2 shows the current list of peptidases that have been purified and characterised from lactococci. A corresponding complement of peptidases has also been found in lactobacilli. A common problem found in these studies, however, is the majority have been carried out using chromogenic substrates or peptides not associated with lactococcal physiology. With the possible exception of the Dipeptidases and the Tripeptidases, little is known about the size of peptide that these peptidases can hydrolyse, and with some exceptions (eg ref 19), their actions on the nutritionally important casein-derived peptides are still largely unknown.

X-Pro Dipeptidyl Aminopeptidase

The first lactococcal peptidase to be cloned and sequenced was the *PepX* gene (30,31). The X-Pro dipeptidyl aminopeptidase that this gene encodes for was considered likely to be one of the essential enzymes associated with the casein hydrolysis due to the proline rich nature of β-casein, comprising 17% of the amino acids. Studies have demonstrated that Proline as a free imino acid does not readily enter the cell and is more likely to be acquired as a peptide by the di/tripeptide transport systems (38). Once these peptides cross the membrane an intracellular prolidase, which can cleave most of the X-Pro dipeptides has been identified in lactococci (32,33). An extracellular enzyme with the ability to form proline-containing dipeptides was therefore anticipated, which acting in concert with the prolidase would supply proline and other amino acids to the cell. The initial purification of this enzyme gave a cell wall location of the enzyme (26), a location also supported to some extent by an immuno-histochemical analysis (39). All other purifications of the enzyme, however, indicate a largely intracellular location, a position supported by analysis of the *PepX* gene. No signal sequence was determined nor was any membrane anchor cross the membrane. Conflicting information exists on the role of this enzyme in the growth of lactococci. In one report (31), a *PepX* deletion mutant in *Lactococcus lactis* subsp. *lactis* NCDO763 was only able to grow in milk at 60% of rate of the wild type, whilst a second report (40) stated a *PepX* deletion mutant did not effect the growth of lactococci in milk.

It should be noted, however, that none of the early formed peptides shown in Table 1 have an X-Pro sequence at the N-terminal and therefore these peptides would not be immediately available as substrates for *PepX*. Other exopeptidases (aminopeptidase) or possibly an endopeptidase are required before these peptides can be substrates for *PepX*.

Table 2 Peptidases isolated from Lactococci

Enzyme[a]	Action	Specificity	Reference
PepN*	X-OOO	Lys, Leu	11-14
PepC*	X-OOO	Lys, Leu, Glu, Phe	15-16
GAP*	X-OOO	Asp, Glu, Ser	17-19
PCP	X-OOO	pGlu	19
DIP*	X-O	Dipeptides (not Pro)	20-22
TRP*	X-OO	Tripeptides (not Pro)	23-25
PepX*	OX-OO	Pro	26-31
PRD	O-X	Pro	32-33
PIP	X-OO	Pro	34
PepO*	...WX-YZ...		35
LEPI	...WX-YZ...		36
LEPII	...WX-YZ...		37

[a]Abbreviation for Peptidases proposed by ref (4)
* Gene cloned and sequenced

Aminopeptidase N (*Pep*N)

PepN is a general aminopeptidase, a metalloenzyme, with a specificity for peptide substrates with N-terminal Lys, Arg and Leu. Analysis of the PepN gene and immunohistochemical studies agree that PepN is an intracellular enzyme (13,14,19,39). Its role in amino acid nutrition and growth is not yet clear. Recently *Pep*N negative mutants were reported not to affect growth in milk (41). A second study showed that the intracellular extract of a PepN negative mutant has an impaired ability to hydrolyse tetra-, penta- and hexapeptide substrates but still retains strong di- and tripeptide hydrolysis capabilities. When grown on a selected medium, limited in the essential amino acid methionine except through a tetrapeptide, the PepN negative mutant did show a reduced growth rate (19). This suggests that the enzyme is involved in oligopeptide processing but, under normal growth conditions in milk, enough essential nutrients can be provided by other proteolytic actions to compensate for the absence of this peptidase activity.

Whilst this enzyme may not be directly essential for growth on milk its involvement in cheese making may be significant. It's ability to hydrolyse hydrophobic amino acids such as leucine, and to a lesser extent phenylalanine, have made it a likely candidate as a debittering enzyme. A study demonstrated that the addition of PepN was able to reduce bitterness in a tryptic digest of β-casein (41). Although *Pep*N has been over expressed in lactococci (14), no studies to date have been reported in cheese using this or *Pep*N negative mutants.

Aminopeptidase C (*Pep*C)

A second general aminopeptidase is also present in some strains of lactococci. *Pep*C is a cysteine type proteinase with a broader range specificity than PepN, active on Lys-, Phe-, His-, Glu-, Leu-β-napthylamides (15). Again, analysis of the *Pep*C gene and the immunohistochemistry show it to be an intracellular enzyme (16,39). Growth studies using PepC-negative strains are underway, but have yet to be reported.

Glutamyl Aminopeptidase (GAP *Pep*A)

A third aminopeptidase, with a specificity for N-terminal glutamate and aspartate containing peptides, has also been described. This metallo enzyme was initially purified and reported to be membrane-bound (17). It appears to have a cell wall/extracellular location from immunohistochemical studies (19). The enzyme, however, has also been purified from an intracellular extract and was shown to have the same N-terminal amino acid sequence as the extracellular form. The gene encoding this sequence has recently been cloned and sequenced and shows no signal or membrane anchoring sequences to support an extracellular location (43). Glutamate is an essential amino acid in lactococci, but the role of this enzyme in its production has yet to be determined. Growth in milk experiments using *Pep*A-negative mutants have not yet been reported.

This enzyme may also have a significant role in development of flavours in ripening cheese. One of the products formed by *Pep*A is glutamate, a recognized flavour enhancer, although its role in cheese flavour development is not clearly understood. Fractionation studies have demonstrated that glutamate is the most prevalent amino acid in the water soluble fraction in matured Cheddar cheese. This fraction is considered to contain the components that make the greatest contribution to the intensity of the flavour (44,45). Cheese trials using *Pep*A-negative starter strains may well give evidence on the role this enzyme plays in flavour development.

Tripeptidase and Dipeptidase

Several peptidases capable of cleaving either dipeptides or tripeptides have been reported (20-24). Lactococci also have a separate Prolidase, a dipeptidase with an X-Pro hydrolysis action (32,33) and an enzyme capable of cleaving di- and tripeptides with N-terminal Proline (34). All are considered to be intracellularly located, with the apparent function of hydrolysing the peptides taken up by the di/tripeptide transport system. The dipeptidase (P.Stroman and F. Mulholland. unpublished data) and the tripeptidase, *Pep*T (46), genes have recently been cloned and sequenced confirming the expected intracellular location of these enzymes. *Pep*T-negative mutants did not show restricted growth in milk (41). No genetic analysis, and consequently no milk growth experiments, have been reported on the prolidase or the proline iminopeptidase.

Endopeptidases

An obvious candidate for further extracellular hydrolysis of the casein derived peptides is an endopeptidase, capable of forming two or more fragments, which would then most likely be a transportable size for one of the peptide transport systems. Several apparently different endopeptidases have now been reported (19, 35-37). Only one, however, the *Pep*O gene, has been cloned and sequenced. The genetic analysis suggests an intracellular location (46), although immunohistochemical studies have suggested a possible extracellular location (39). In milk growth experiments *Pep*O-negative mutants did not show any differences compared to the wild type (41). The function of an intracellular endopeptidase in the hydrolysis of casein peptides for growth is debatable. Lactococci already have several general intracellular peptidases present, (*Pep*N, *Pep*C, DIP, TRP and PRD), that are capable of hydrolysing the 2-6 amino acid peptides thought to be able to enter via the peptide transporters.

5. CURRENT POSITION ON PEPTIDASES.

At our current level of knowledge, a number of observations relating to the ability of the lactococcal peptidases to utilise casein-derived peptides and their role in growth have yet to be fully explained.

It has generally been accepted that the caseins are the major source of amino acids utilised by the organisms for growth in milk and the early forming casein-derived peptides are considered the most likely candidates to provide the nutrients required by the organisms during growth. Current knowledge, however, on the peptide transport systems of lactococci suggests that they should not be able to transport the peptides produced from the action of the extracellular cell wall associated proteinase on the caseins. Lactococci have two distinct peptide transport systems, a proton motive force dependent di/tripeptide transporter, and an ATP driven oligopeptide transporter. From the characterisation of the latter, the maximum size peptide shown to be transported is 6 amino acids (47). *In vitro* studies on the formation of the peptides by the action of the proteinase on caseins have shown very few small peptides of less than 7 amino acids. The genes for both the oligopeptide and the di/tripeptide transport system have now been sequenced. In milk growth experiments, using mutants (derived by either chemical or genetic manipulations) deficient in either of these two peptide transport systems, showed that both are essential (48,49). It is generally thought that some further hydrolysis of the peptides needs to occur extracellularly if these peptides are to be utilised for growth.

This raises the second question on the involvement of lactococcal peptidases in growth. What is the true physiological location of these peptidases? To date, the evidence for the extracellular location of any of the peptidases is conflicting. The immunohistochemical evidence for *Pep*X, *Pep*O and *Pep*A indicates that these enzymes are cell wall/membrane associated (19,39), and indeed, some have been purified from cell wall/membrane preparations (17,19,26). These same enzymes, however, have also been purified from intracellular extracts and all the genetic evidence suggests they are intracellular. The possibility of some unknown mechanism of translocation of the enzymes should not yet be ruled out. Currently there two other candidates as extracellular peptidases, a 36 kDa aminopeptidase isolated from the cell wall of *Streptococcus cremoris* AC1 in 1985 (50), and the recently purified 23 kDa cell wall peptidase from *Lactococcus lactis* subsp. *cremoris* IMN-C12 (25). Surprisingly, no further information on the first of these enzymes is available, despite being one of the first lactococcal peptidases to be described.

To date no single peptidase has been shown to be an essential enzyme when lactococci are grown in milk. Some studies using double peptidase-negative mutants have also been performed but still no difference in growth rates and acidification could be found (41). With the wide variety of general, rather than highly specific, peptidases found intracellularly this is perhaps not surprising and even triple and quadruple peptidase-deficient mutants are required. An example of this can be demonstrated by the potential routes for the formation of glutamate, one of the essential amino acids by peptidolysis. Lactococci possess several enzymes that are capable of hydrolysing peptides to produce glutamate. PepA (17-19), PepC (15), the dipeptidase (22) and the tripeptidase (23,24) all show activity against glutamate-containing (peptides or chromogenic) substrates. In this case, it may be necessary, to construct a quadruple peptidase-negative mutant before any restriction of growth is observed.

6. FUTURE AREAS OF RESEARCH

As noted above, one of the next areas of research will undoubtedly be the development of the multiple peptidase-negative strains, aimed at removing the ability of the lactococci to hydrolyse peptides for the production of one of the essential amino acids. A word of caution should, however, be noted. It is generally assumed that these peptidases are involved in the nutritional role of supplying the lactococci with amino acids for growth. Some of these enzymes may also have a significant metabolic role not related to casein peptide hydrolysis.

When using these peptidase-negative strains, restricted growth could be due to this dysfunction, rather than an inability to hydrolyse casein peptides. Highly autotrophic organisms, which do not require a proteolytic system to supply their amino nitrogen nutrients, also contain significant peptidase activities (G.W. Niven, personal communication). Other metabolic functions for the peptidases in lactococci should not be discounted.

Another issue that has yet to be addressed in peptidases is genetic variance. The lactococcal cell wall associated proteinase clearly has several natural genetic variants. These variants have some different cleavage properties on casein which are considered important in cheese manufacture (51). Whether or not genetic variation occurs in peptidases has not yet been examined. An example of this could be the ability of the lactococcal dipeptidase to cleave N-terminal glutamate dipeptides. The dipeptidase isolated from *Lactococcus lactis* subsp. *cremoris* Wg2 does not cleave Glu-Ala (21), whilst a similar dipeptidase from *Lactococcus lactis* subsp. *lactis* NCDO 712 is able to cleave this and other glutamate-containing dipeptides (22). To address this issue, more substantial specificity studies, using standardised conditions, and in some cases with more natural peptide substrates, allowing comparisons of individual peptidase specificities between different (industrially important?) strains are required. Further studies at the genetic level may then show the rationale for these specificity differences. Linked to this are the levels of different peptidase activities found in individual strains. Little work has yet been done comparing the amount of enzyme found in different strains.

A third area for future study is regulation of expression of the peptidases. Again, the lactococcal proteinase is known to be subject to regulation of its expression in selected peptide-containing media (52). Whether or not any of the peptidases are subject to regulation has not been addressed.

7. CONCLUSIONS

In recent years, a substantial advance has been made in the knowledge of peptidases in lactococci, both biochemically and genetically. The physiological role of the individual peptidases, however, relating to their involvement in secondary hydrolysis of casein peptides and growth remains largely unanswered.

Acknowledgements

This work is supported in part by the EC BRIDGE T-Project on the Biotechnology of Lactic Acid Bacteria. Contract No. BIOT-CT91-0263.

REFERENCES

1. B.A. Law and J. Kolstad, Antonie van Leeuwenhoek, 1983, 49, 225.
2. J. Kok, FEMS Microbiol. Rev., 1990, 87, 15.
3. G.G. Pritchard and T. Coolbear, FEMS Microbiol. Rev., 1993, 12, 179.
4. P.S.T. Tan, B. Poolman and W.N. Konings, J. Dairy Res., 1993, 60, 269.
5. R.C. Lawrence, T.D. Thomas, and B.E. Terzaghi, J. Dairy Res., 1976, 43, 141.
6. J. Kok, J.M. van Dijl, J.M.B.M. van der Vossen and G. Venema, Appl. Environ. Microbiol., 1985, 50, 94.
7. F.A. Exterkate and G.J.C.M. De Veer, Appl. Microbiol. Biotechnol., 1987, 25, 471.
8. E.J. Smid, B. Poolman and W.N. Konings, Appl. Environ. Microbiol., 1991, 57, 2447.
9. V. Monnet, D. Le Bars and J.-C. Gripon, FEMS Microbiol. Lett., 1986, 36, 127.
10. J.R. Reid, C.H. Moore, G.G. Midwinter and G.G. Pritchard, Appl. Microbiol. Biotechnol., 1991, 35, 222.
11. P.S.T. Tan and Konings, Appl. Environ. Microbiol., 1990, 56, 526.
12. F.A. Exterkate, M. DeJong, G.J.C.M De Veer and R. Baankreis, Appl. Microbiol. Biotechnol., 1992, 37, 46.
13. P. Stroman, Gene, 1992, 113, 107.
14. I.J. Van Alen-Boerrigter, R. Baankreis and W.M. De Vos, Appl. Environ. Microbiol., 1991, 57, 2555.
15. E. Neviani, C.Y. Boquien, V. Monnet, L.P. Thanh and J.-C. Gripon, Appl. Environ. Microbiol., 1989, 55, 2308.
16. M.-P. Chapot-Chartier, M. Nardi, M.-C. Chopin, A. Chopin and J.-C. Gripon, Appl. Environ. Microbiol., 1992, 59, 330.
17. F.A. Exterkate and G.J.C.M. De Veer, Appl. Environ. Microbiol., 1987, 53, 577.
18. G.W. Niven, J. Gen. Microbiol. 1991, 137, 1207.
19. R. Baankreis, PhD Thesis, University of Amsterdam, 1992.
20. I.K. Hwang, S. Kaminogawa and K. Yamauchi, Agric. Biol. Chem., 1981, 45, 159.
21. A. Van Boven, P.S.T Tan, and W.N Konings, Appl. Environ. Microbiol., 1988, 54, 43.
22. S. Movahedi and F. Mulholland, 1993, submitted for publication.
23. B.W. Bosman, P.S.T. Tan, and W.N. Konings, Appl. Environ. Microbiol., 1990, 56, 1839.
24. C.L. Bacon, M. Wilkinson, P.V. Jennings, I. Ni Fhaolain and G. O'Cuinn, Int. Dairy J., 1993, 3, 163.
25. S. Sahlstrom, J. Chrzanowska, T. Sorhaug, Appl. Environ. Microbiol., 1993, 59, 3076.
26. B. Kieffer-Partsch, W. Bockelmann, A. Geis and M. Teuber, Appl. Microbiol. Biotechnol., 1989, 31, 75.
27. C. Zevaco, V. Monnet and J.-C. Gripon, J. Appl. Bacteriol., 1990, 68, 357.

28. M. Booth, I. Ni Fhaolain, P.V. Jennings, and G. O'Cuinn, J. Dairy Res., 1990, 57, 89.
29. R.J. Lloyd and G.G. Pritchard, J. Gen. Microbiol. 1991, 137, 49.
30. B. Mayo, J. Kok, K. Venema, W. Bockelmann, M. Teuber, H. Reinke and G. Venema, Appl. Environ. Microbiol., 1991, 57, 38.
31. M. Nardi, M.-C. Chopin, A. Chopin, M.-M. Cals and J.-C. Gripon, Appl. Environ. Microbiol., 1991, 57, 45.
32. S. Kaminogawa, N. Azuma, I.K. Hwang, Y. Susuki and K. Yamauchi, Agric. Biol. Chem., 1984, 48, 3035.
33. M. Booth, P.V. Jennings, I. Ni Fhaolain and G. O'Cuinn, J. Dairy Res., 1990, 57, 245.
34. R. Baankreis, and F.A. Exterkate, Syst. Appl. Microbiol. 1991, 14, 317.
35. P.S.T. Tan, K.M. Pos and W.N. Konings, Appl. Environ. Microbiol., 1991, 57, 3593.
36. T.-R. Yan, N. Azuma, S. Kaminogawa and K. Yamauchi, Appl. Environ. Microbiol., 1987, 53, 2296.
37. T.-R. Yan, N. Azuma, S. Kaminogawa and K. Yamauchi, Eur.J. Biochem., 1987, 163, 259.
38. E.J. Smid and W.N. Konings. J. Bacteriol., 1990, 172, 5286.
39. P.S.T. Tan, M.-P. Chapot-Chartier, K.M. Pos, M. Rousseau, C.-Y. Boquien, J.-C. Gripon and W.N. Konings, Appl. Environ. Microbiol., 1992, 58, 285.
40. B. Mayo, J. Kok, K. Venema, W. Bockelmann, A.J. Haandrickman, K.J. Leenshout and G. Venema, Appl. Environ. Microbiol., 1993, 59, 2049.
41. I. Mierau, A.J. Haandrikman, K.J. Leenshout and G. Venema, FEMS Microbiol. Rev., 1993, 12, P83.
42. P.S.T. Tan, PhD Thesis, University of Groningen, 1992.
43. K. I'Anson, H. Griffin, S. Movahedi, F. Mulholland and M. Gasson. FEMS Microbiol. Rev., 1993, 12, P74.
44. J.W. Aston and L.K. Creamer, New Zealand J. Dairy Sci.Technol., 1986, 21, 229.
45. A.J. Cliffe, J.D. Marks and F. Mulholland, Int. Dairy J., 1993, 3, 379.
46. I. Mierau, A.J. Haandrikman, K.J. Leenshout, P.S.T. Tan, J. Kok, W.N. Konings and G. Venema, FEMS Microbiol. Rev., 1993, 12, P84.
47. E.R.S. Kunji, R. Plapp, B. Poolman and W.N. Konings, FEMS Microbiol. Rev., 1993, 12, P62.
48. E.J. Smid, R. Plapp and W.N. Konings, J. Bacteriol., 1989, 171, 6135.
49. S. Tynkkynen, G. Buist, E.R.S. Kunji, A.J. Haandrikmann, J. Kok, B. Poolman and G. Venema, FEMS Microbiol. Rev., 1993, 12, P44.
50. A. Geis, W. Bockelmann and M. Teuber, Appl. Microbiol. Biotechnol., 1985, 23, 79.
51. F.A. Exterkate and G.J.C.M. De Veer, Syst. Appl. Microbiol.,1989, 11, 108.
52. F.A. Exterkate, J. Dairy Sci., 1985, 68, 562.

Functional Milk Protein Products

Daniel M. Mulvihill

FOOD CHEMISTRY DEPARTMENT, UNIVERSITY COLLEGE, CORK, IRELAND

1 INTRODUCTION

Bovine milk contains only ~ 3.5% protein. However, the heterogeneity of the protein and the range of processes used for its recovery makes milk protein products perhaps the most versatile functional food protein products available.

The heterogeneity of bovine milk proteins has been well documented[1]; the protein falls into two main categories based on solubility at pH 4.6 at > ~ 8°C. Under these conditions ~ 80% of the total nitrogen precipitates and is referred to as casein while 20% remains soluble in the serum or whey; ~ 15% being whey protein with the remainder being non-protein nitrogenous compounds.

Casein consists of four principal primary proteins (gene products), α_s1, α_{s2}, ß and κ in the approximate ratio 40:10:35:12 and several minor proteins, most of which are present as a result of post secretion proteolysis of the primary caseins by indigenous milk enzymes.

The primary caseins also exhibit microheterogeneity due to variations in the degree of phosphorylation or glycosylation, disulphide-linked polymerization and genetically-controlled amino acid substitution.

The whey proteins include ß-lactoglobulin (ß-lg), α-lactalbumin (α-la), bovine serum albumin (BSA) and immunoglobulins (IgG, IgA, IgM), which represent 50, 20, 10 and 10%, respectively, of total whey protein in bovine milk. Whey also contains a number of minor proteins, including proteose-peptones, a heterogenous mixture of heat stable, acid soluble (pH 4.6) polypeptides, precipitated by 12% trichloroacetic acid, most of which are the result of proteolysis of the caseins by indigenous milk enzymes; biologically active proteins such as lactoperoxidase and lactotransferrin; casein glycomacropeptide, the sequence of κ-casein from residues 105 to 169 that is generated by hydrolysis of κ-casein by chymosin and is present in rennet casein whey and some cheese wheys; and several enzymes. The properties of the caseins and whey proteins differ very significantly:

i. The caseins are insoluble at their isoelectric points (~ pH 4.6) while in the ionic environment of milk the whey proteins are soluble at their isoelectric points (~ pH 5.0).

ii. Addition of crude proteinases preparations, known as rennets, to milk induces limited proteolysis of casein and results in its coagulation while the whey proteins remain soluble.

iii. Caseins are extremely heat stable while the whey proteins are heat labile. On heating milk at > 72°C whey proteins become denatured and interact with caseins to form a complex.

iv. In milk the caseins occur as spherical, macromolecular complexes with molecular weights of ~ 10^8 and mean diameters of ~ 100 nM, known as micelles, that also contain inorganic salts, principally calcium, phosphate, magnesium and citrate, referred to collectively as colloidal calcium phosphate (CCP), while the whey proteins are in solution normally in monomer or dimer form.

These differences between casein and whey proteins are exploited in industrial methods used for the recovery of functional milk protein products.

This paper will give an overview of processes used for the production of a range of functional, dehydrated milk protein-enriched products.

PRODUCTION OF CASEINS AND CASEINATES

Conventional Methods for Production of Caseins

Caseins are manufactured from skim milk as this ensures that the fat content is low enough to minimize flavour defects during subsequent storage of the dried products. The first step in the process is the destabilization of the casein to render it insoluble. This may be achieved by:

a. inoculating skim milk with a mixed or multiple defined-strain starter and incubating at 22 to 26°C for 14-16 h; the added starter slowly ferments some of the lactose to lactic acid and a casein gel network or coagulum is formed as the pH of the milk falls slowly under quiescent conditions to the isoelectric pH of the casein.

b. spraying dilute (1-2 M) mineral acid, usually HCl, under pressure into milk (preheated to 25-30°C) flowing in the opposite direction to give a precipitation pH of ~ 4.6.

c. mixing skim milk at < 10°C with a cation exchange resin in the hydrogen form in a reaction column; this replaces cations in the milk by H^+ to give a pH of ~ 2.2, this acidified milk is then mixed with untreated milk to give the final desired precipitation pH of ~ 4.6.

d. adding any of a number of proteinases (rennets) which can coagulate milk at its natural pH (~ 6.7) in a two stage process; the first stage involves the specific hydrolysis of κ-casein to yield para-κ-casein and (glyco)macropeptides, while the second stage involves coagulation of the rennet-altered casein micelles by Ca^{2+} at temperatures of about 30°C.

The casein destabilized milk is then heated to temperatures in the range 50-60°C by direct steam injection and held for a period of about 1 min to ensure complete coagulation and agglomeration of the curd prior to separation of the curd and whey. The first three destabilization processes result in the production of an acid curd and the caseins produced are referred to as lactic acid, hydrochloric acid and ion exchange caseins, respectively. The latter destabilization process results in the production of a high pH curd referred to as rennet casein and this curd retains a high ash content, especially colloidal calcium phosphate (CCP), calcium and phosphate.

Following destabilization, the insoluble casein is separated from the soluble whey proteins, lactose and salts, washed to remove residual soluble solids and dried. Traditional driers used are semi-fluidized, vibrating type driers or pneumatic ring driers. A drying technique, referred to as 'attrition drying', based on the principle of grinding and drying in a single operation, is also widely used in casein plants since it allows the production of a casein product closely resembling spray dried casein.

Dried casein is relatively hot as it emerges from the drier and the moisture content of individual particles varies. Therefore, it is necessary to temper or cool and blend the dried curd prior to milling in roller or pin-disc mills to produce casein particles of the size range required by the end-users of the products.

Non-conventional Methods for Production of Caseins

The caseins can be precipitated from milk at its normal pH by addition of ethanol to about 40%; the concentration of ethanol necessary to induce precipitation decreases to about 10-15% as the pH is reduced to ~ 6.0. The use of ethanol to coagulate casein for industrial preparation has been described[2]. An alternative method to the direct addition of ethanol to milk to induce casein precipitation is the use of ethanol to dissolve the lactose from dry skim milk powder leaving an insoluble residue which may be regarded as a total milk protein[3].

When milk or milk concentrate prepared by ultrafiltration is frozen and stored at ~ -10°C, the casein micelles are cryo-destabilized and precipitate when the milk is thawed. The potential of this method for the production of casein has also been investigated[4,5]. When cryodestabilized casein is dispersed in water at 40°C it retains micellar properties.

Most of the casein micelles in milk may be sedimented by ultracentrifugation at > 100,000 g for > 1 h. This method is used to prepare casein micelles for research purposes but is not suitable for industrial scale production of casein micelles. However, a new process being developed for the production of "native" casein[6,7] involves electrodialysis of skim milk at 10°C against acidified whey to reduce the pH to ~ 4.5 or ~ 2.8; in the acidified milk the casein is reported to be in a "metastable" state and is capable of being physically separated by low centrifugal forces. The recovered "native" casein is washed with demineralized water and finally recovered by ultrafiltration/diafiltration. The product is claimed to have properties similar to those of native casein micelles.

With fractionating membranes of suitable pore size and characteristics it should be possible to fractionate casein micelles and whey proteins. This process is referred to as microfiltration as the membranes have nominal cut-offs in the range 0.1 to 10 μM. The application of membranes of this type for the recovery of casein micelles is being explored[8].

Caseinate Production

Acid precipitated caseins are insoluble in water but wet acid casein curd or dry acid casein will dissolve in alkali under suitable conditions to yield water-soluble caseinates. Sodium caseinates, prepared by solubilizing acid casein with NaOH followed by spray drying, is the water-soluble casein most commonly used in foods. However, other bases and processing conditions may be used to produce a wide range of different caseinates[9].

Fractionation of Casein

As previously stated, bovine milk contains 4 primary caseins, α_{s1}-, α_{s2}-, β- and κ-; it has been possible for many years to fractionate these proteins on a laboratory scale based on differences in solubility in urea solutions at acid pH values or by selective precipitation with $CaCl_2$. It is also possible to fractionate the caseins by various forms of chromatography. Obviously, these methods are not very amenable to scale-up for industrial application because of the high costs associated with these fractionation methods.

There are a number of incentives for developing low cost methods for the fractionation of caseins on an industrial scale, e.g.

1. β-Casein has exceptionally high surface activity and may find special applications as an emulsifier or foaming agent.

2. Human milk contains β- and κ-caseins but no α-caseins;

hence, β-casein should be an attractive ingredient for bovine milk-based infant formulae.

3. κ-Casein is responsible for the stability of casein micelles and if available in sufficient quantities might be a useful additive for certain milk products.

4. All the caseins, especially β-casein, and indeed all milk proteins, contain amino acid sequences which have biological activities when released by proteolysis; the best studied of these are the β-caseinomorphines. The preparation of biologically active peptides will require purified proteins.

A number of methods for fractionating casein into β-casein-rich and $α_s$-/κ-casein-rich fractions on a potentially industrial scale have been developed. These methods exploit the ionic strength and/or temperature dependent association characteristics of the caseins. β-Casein is the most hydrophobic of the caseins and undergoes strong temperature dependent association; at 4°C it exists in monomeric form but it associates strongly as the temperature increases. This characteristic was exploited by Allen et al[10]. to prepare β-casein by renneting calcium caseinate at 4°C; under these conditions, β-casein remains soluble while $α_s$- and para-κ-caseins coagulate. A method for the isolation of β-casein by microfiltration of milk or calcium caseinate at 5°C was developed by Terre et al.[11] Famelart et al.[12] optimized the same technique to purify β-casein from whole casein at 4°C and pH 4.2-4.6. Murphy and Fox[13] developed a method for the fractionation of a dilute sodium caseinate solution by ultrafiltration into a β-casein-rich permeate and an $α_s$-/κ-casein-rich retentate. The β-casein was recovered from the permeate by raising the temperature to 40°C and recovering the associated protein by UF.

Production of Whey Protein Enriched Products

Whey is the serum or liquid portion remaining after removal of the curd formed by rennet-type enzymes and/or acid from milk during the manufacture of cheeses and caseins. Composition of whey varies depending on milk composition, cheese variety, casein type and processing conditions used in the manufacture of the cheese or casein. There are two principal types of whey; sweet whey (minimum pH 5.6) from the coagulation of milk by rennet-type enzymes in the manufacture of several cheeses and rennet casein, and acid whey (maximum pH 5.1) from the coagulation of milk primarily with acid in the manufacture of acid casein and acid cheeses. Wheys typically contain ~ 50% of the total solids of the original milk, including about 20% of total milk proteins, almost all the milk sugar, lactose, and a variable proportion of the milk salts. The different mineral compositions and pH values are important factors for consideration in selecting processing conditions and the different mineral contents may have significant effects on the functionality of dried whey and whey protein-enriched products.

Traditionally, whey was regarded by the dairy industry as an undesirable by-product, of low commercial value, which presented the industry with problems of disposal. However, the imposition of strict controls on the disposal of waste necessitated the construction and operation of effluent treatment plants, which proved costly for the industry. This, together with a recognition that whey was a potentially valuable source of nutrients, stimulated interest in the development of commercially viable processes to convert liquid whey into valuable products, suitable for use in both human and animal foods.

Both acid and rennet whey types contain 0.65-0.8% (w/v) protein which corresponds to 10-12% (w/w) protein on a dry basis; the chief constituent is lactose, which represents approximately 70% of the solids[14]. Acid whey contains a higher concentration of minerals than rennet whey due to dissolution of the colloidal calcium phosphate component of the casein micelles during acidification. Cheese whey derived from whole milk has higher concentrations of residual milk fat than that derived from skim milk but the residual milk fat is also a function of clarification, separation and other pre-treatments employed to remove this component from the whey prior to protein recovery.

During whey processing, whey and whey protein-enriched solutions are usually pasteurized and thermally concentrated at minimum temperatures and holding times and maintained at low temperatures during storage to minimize microbial and physico-chemical deterioration of the proteins and other whey constituents that would adversely alter the functional and organoleptic properties of the resulting protein-enriched products.

A brief review of the the various methods known to be in commercial or advanced pilot scale use for the recovery of whey proteins follows:

Whole Whey Powder

Liquid whey may be dehydrated per se to produce a powder with a protein content of 10-12% (w/w). It is usual to clarify the whey to remove residual cheese curd or casein particles and/or to separate the residual fat (if the whey is derived from cheese manufacture) prior to dehydration. After clarification and/or separation, the whey is usually concentrated to 50-60% total solids. Concentration is carried out under vacuum at temperatures below 70°C to avoid protein denaturation. Reverse osmosis may be employed to concentrate the whey to approximately 25% total solids to increase the capacity of the evaporator[15]. Drying of the concentrate is achieved either by spray drying or roller drying, though the latter is increasingly rare due to greater heat damage to the powder.

Special methods have been developed to produce non-hygroscopic, non-caking and wettable whey powders. Following concentration, the whey concentrate may be cooled quickly to 28-30°C, seeded with lactose nuclei and then further cooled slowly over a period of hours to 15-18°C to crystallize up to 80% of the lactose as α-monohydrate-lactose, which is less hygroscopic than ß-anhydrous-lactose. The concentrate is then spray dried in a single stage process to produce a dense powder with small agglomerates, or in a two stage process to produce a coarse agglomerated powder. In the two stage process, which is more applicable to acid wheys, low temperatures are used in the first stage of drying to give a product containing ~ 12% moisture. The intermediate product is held for 3-5 min while more lactose crystallizes as the α-monohydrate before drying in a secondary drier[16].

Demineralized Whey Powder

The high ash content of whole whey powder can adversely affect the flavour and nutritional quality of the product. Thus, electrodialysis and/or ion-exchange processes may be used to demineralize the whey concentrate prior to drying. Demineralization is particularly useful when applied to sweet whey; powders with a very low mineral content, suitable for use in baby and special diet foods can be produced.

Electrodialysis has been shown to be cost-effective for up to 70% mineral removal from cheese whey. Ion-exchange processes are suitable for removal of up to 90% of the minerals of whey. Many modern whey demineralization plants use a combination of electrodialysis and ion-exchange processes.

Demineralized, Delactosed Whey Powder

Lactose removal is usually accomplished by crystallization. Typically, the whey is concentrated by evaporation to 40-60% total solids, the concentrate slowly cooled and seeded with lactose nuclei to induce lactose crystallization, the mother liquor (delactosed whey) separated from the lactose crystals by decanting off, or by centrifugation, and dried to produce delactosed whey powder containing about 25% protein on a dry basis. However, delactosed whey powder has a high mineral content (up to 25%) which restricts its use in foods. Therefore, it is common to demineralize whey in conjunction with lactose removal. In a typical process for the production of demineralized, delactosed whey powder, pasteurized whey or whey concentrated to 20-30% total solids is adjusted to pH 6.2-6.4, clarified and demineralized by electrodialysis or a combination of electrodialysis and ion exchange processes. The demineralized whey is concentrated to 40-60% total solids by evaporation at temperatures less than 70°C (to avoid protein denaturation). The concentrate is cooled in a controlled manner to induce lactose crystallization. The lactose is recovered by centrifugation and the mother liquor may be concentrated and spray dried to produce a powder containing 25-35% protein or may be further processed to products containing higher protein contents.

Alternatively, whey can be delactosed before demineralization. In this process whey is concentrated by evaporation to 50-60% total solids, cooled to induce lactose crystallization, 40-60% of the lactose is removed by centrifugation. Residual lactose crystals in the mother liquor are dissolved by heating to 43-50°C. The mother liquor is clarified to remove insoluble protein and then demineralized by electrodialysis and/or ion exchange processes at up to 33% total solids to remove 20-55% of the minerals.

Whey Protein Concentrate (WPC) and Whey Protein Isolate (WPI)

The terms whey protein concentrate (WPC) and whey protein isolate (WPI) refer to whey protein powders containing 35-95% protein on a dry basis. These products are usually prepared on a commercial scale by such processes as ultrafiltration, diafiltration and ion exchange chromatography.

Ultrafiltration-Diafiltration

Ultrafiltration (UF) is a pressure membrane filtration process that facilitates the selective separation of whey proteins from lactose, salts and water under mild conditions of temperature and pH. It is a physico-chemical separation technique in which a pressurized solution flows over a porous membrane that allows the passage of only relatively small molecules. The retained solution (retentate) flows over the membrane, while under the influence of pressure water flows through the membrane, together with low molecular weight solutes (the permeate). The protein is retained by the membrane and is therefore concentrated relative to other solutes in the retentate. Fat globules and suspended solids are also retained.

The membranes used in UF are asymmetric microporous structures, the effective layers of which contain pores with diameters ranging from 1 to 20 nm. Commonly used membrane configurations include tubular, spiral-wound, plate and frame and hollow-fibre, with each configuration offering advantages and disadvantages for particular applications. The membranes are manufactured from synthetic polymers (e.g. polysulphone or polyamide). They are characterized by high resistance to high temperatures (up to 100°C), can withstand a wide pH range (1-13) and can be cleaned with agents normally used in the dairy industry (e.g. HNO_3 and NaOH). Although UF is currently the method of choice for the commercial manufacture of WPC of varying protein concentration, it has several major problems in operational performance. These problems include: high capital and operating costs; membrane fouling, with concomitant loss of permeate flux rate; incomplete removal of low molecular weight solutes unless diafiltration (dilution of retentate with water and repeated UF) is used; cleaning, sanitation and related microbial problems; disposal of large volumes of permeate.

Prior to processing, whey is commonly pre-treated by methods involving pH and/or temperature adjustments, addition of calcium or calcium complexing agents and either quiescent standing, centrifugation or microfiltration to dissolve colloidal calcium phosphate and/or to remove insoluble cheese curd or casein fines, milk fat and calcium lipophosphoprotein complexes[17-25]. These pre-treatments increase flux during ultrafiltration, prevent fouling of the membranes and modify the properties of the whey concentrates.

Because of the variation in feed rate throughout the time of UF (initially the flux is high, falling quickly due to concentration polarization and then more slowly as the membrane fouls), it is usually necessary to build up a buffer stock of whey before UF commences. The UF process is generally operated at temperatures above 50°C to minimize microbial growth. Most modern UF plants are continuous multiple stage in series recirculation systems which enables the protein to total solids ratio in the retentate to be readily controlled. The maximum ratio which can be achieved by UF on modern plants is about 0.65:1.0; above this value the viscosity of the retentate is such that the flux rate becomes excessively low. However, higher ratios may be achieved by the addition of water to the retentate at the final stages of the process. The water dilutes the retentate, decreases the viscosity and, as it permeates, washes out lactose and minerals. This process is called diafiltration and is employed to achieve ratios of protein to total solids up to ~ 0.80:1.

In most commercial plants the final retentate is relatively dilute and some further concentration is required before spray drying. Low temperature evaporation (~ 40°C) to 15-20% (w/v) protein in the concentrate is the normal commercial practice. Drying results in the production of powders containing 35-80% (w/w) protein.

Ion Exchange Adsorption
Whey proteins are amphoretic molecules and therefore acquire a net charge, depending on the pH. At pH values lower than their isoelectric point (pH ~ 4.6), whey proteins have a net positive charge and behave as cations which can be adsorbed on cation exchangers. At pH values above their isoelectric point, the proteins have a net negative charge and behave as anions which can be adsorbed on anion exchangers. Media with suitable pore sizes and surface characteristics have been developed specifically for the recovery of proteins from dilute solutions, depending upon the pH of the medium. Two major ion exchange fractionation processes have been commercialized for the manufacture of WPI.

The "Vistec" process uses a cellulose-based exchanger in a stirred tank reactor[26,27]. The process involves a series of steps that are performed as a fractionation cycle: (1) whey is adjusted to pH < 4.6 with acid, pumped into a tank reactor and stirred to allow protein adsorption onto the ion exchanger, (2) lactose and other unadsorbed materials are filtered off, (3) the resin is resuspended in water and the pH adjusted to > 5.5 with alkali to release the proteins from the ion exchanger, (4) the aqueous solution of proteins is separated from the resin by filtration in the tank reactor, concentrated by ultrafiltration and evaporation and spray dried as WPI containing ~ 95%

protein. UF treatment of the protein-rich eluate fraction is essential for purification and concentration of the protein.

The "Spherosil" processes[28,29] use either cationic Spherosil S or anionic Spherosil QMA ion exchangers and fractionation is accomplished in fixed-bed column reactors. Acidified whey at pH < 4.6 is applied to the Spherosil S column reactor to allow protein adsorption by the strongly acidic cation resin. After lactose and other unadsorbed solutes have been eluted, the pH is raised by addition of alkali to elute adsorbed proteins from the reactor. The protein-rich eluate fraction is concentrated by UF and evaporation and spray dried to produce WPI. Sweet whey at pH > 5.5 is applied to the Spherosil QMA column reactor to permit negatively charged protein molecules to be adsorbed by the strong anionic ion exchanger. After elution of non-protein materials, the proteins are released by lowering the pH with acid. Released proteins are concentrated and spray dried as WPI, as for the Spherosil S process.

These adsorption processes recover ~ 85% of the protein under ideal operating conditions and the recovered concentrates are characterized by high protein and low lactose and lipid concentrations and have good functionality. However several major problems are associated with these ion exchange processes including: (1) production of large volumes of rinse water, regeneration solutions and deproteinized whey that must be processed or disposed of, (2) the need to concentrate and purify the dilute protein-containing eluate by UF, evaporation and drying, (3) the excessive time requirement for conducting each fractionation cycle, (4) microbial contamination of the reactor.

Lactalbumin Production

Whey proteins are globular proteins and are readily denatured on heating. On transformation from their globular conformations to more random structures, sulphydryl and hydrophobic groups are exposed and protein-protein interactions occur. The extent of aggregation and precipitation of the denatured proteins depends on heating temperature and holding time, pH and concentration of calcium. Commercial precipitation conditions employed to recover heat denatured protein depend on whey type and the desired final product characteristics and whey may be pre-concentrated and/or demineralized prior to precipitation[30]. The precipitated protein, referred to as lactalbumin, may be recovered by settling and decanting, vacuum filtration, self-desludging centrifuges or horizontal solid-bowl decanters. The precipitate may be washed to reduce mineral and lactose contents and dried in spray, roller, ring or fluidized bed driers. Protein yields may be up to 80% of that in the whey and lactalbumin containing up to 90% protein on a dry weight basis may be recovered, depending on precipitation pH and degree of washing.

Fractionation of Whey Proteins

Techniques for the isolation of individual whey proteins on a laboratory scale by salting-out, ion exchange chromatography and/or crystallization have been available for about 40 years. Owing to the unique functional, physiological or other biological properties of many of the whey proteins, there is an economic incentive for their isolation on an industrial scale.

For example, ß-lg, the principal whey protein in bovine milk, produces better thermo-set gels than α-la. However, human milk does not contain ß-lg which is the most allergenic of the bovine milk proteins for the human infant; therefore, α-la would appear to be a more appropriate protein for the preparation of humanized baby formulae than total whey protein.

ß-Lactoglobulin and α-Lactalbumin

A number of methods have been developed for the separation of α-la and ß-lg. Probably the most commercially amenable of these are methods in which the low heat stability of apo-α-la is exploited to precipitated it from whey, leaving ß-lg, BSA and Ig in solution[23,31-33]. α-La is a Ca-containing metalloprotein which is denatured at relatively low temperatures but renatures on cooling. The protein loses its Ca and is transformed into the apo form on acidification to ≤ pH 5. In the apo form it aggregates on heating to 55°C and can be removed by centrifugation, filtration or microfiltration.

α-la and ß-lg are insoluble in pure water at their isoelectric points; ß-lg requires a higher ionic strength for solubility than α-la, a characteristic which may be exploited to fractionate α-la and ß-lg[34]. Whey is concentrated by UF, acidified to pH 4.65 and demineralized by electrodialysis to < 0.023% ash; ß-lg precipitates and may be recovered by centrifugation with a yield of > 90%.

The principal whey proteins may also be fractionated by $FeCl_3$ at pH 3[35], sodium hexametaphosphate[36], NaCl[37] or TCA[38].

The ion exchangers used to recover WPI may also be used to fractionate whey proteins[39]. All the whey proteins are adsorbed initially on Spherosil QMA but on continued passage of whey through the column, ß-lg, which has a higher affinity for this resin than the other proteins, displaced α-la and BSA giving a mixture of these proteins in the eluate; a highly purified ß-lg can be obtained by eluting the protein-saturated column with 0.1 M HCl.

At a pressure of 2000 atmospheres the ß-lactoglobulin in bovine whey was denatured and completely digested by thermolysin while there was no effect on α-lactalbumin[40]. This was proposed as a method of preparing an enzymatic whey hydrolysate to simulate human milk protein.

Minor Whey Protein Products

Whey contains a number of proteins that are of biological

or pharmaceutical interest. Many of these proteins may eventually find commercial application as isolation procedures are improved but at present, 4 are of commercial interest, viz. lactoperoxidase, lactotransferrin, immunoglobulins and glycomacropeptide.

Lactoperoxidase

Lactoperoxidase (LPO) is a broad specificity peroxidase present at high concentrations in bovine milk but at low levels or not at all in human milk.

LPO has attracted considerable interest since it has been shown to be involved in the antibacterial activity of various secretions. In milk the antibacterial system consists of LPO, H_2O_2 and ^-SCN. The active species is hypothiocyanate ($OSCN^-$) or some higher oxidation species. Milk normally contains adequate LPO activity and some ^-SCN (the concentration depends on the animal's diet) but no indigenous H_2O_2; therefore, to activate the indigenous antibacterial system, H_2O_2 must be added or produced in situ, e.g. by the action of glucose oxidase or xanthine oxidase and it is usually necessary to supplement the indigenous ^-SCN.

Commercial interest in LPO involves: (1) activation of the indigenous enzyme for cold sterilization of milk or in the mammary gland to protect against mastitis, (2) addition of isolated LPO to calf or piglet milk replacers to protect against enteritis, especially when the use of antibiotics in animal feed is not permitted.

LPO is positively charged at neutral pH and can be isolated from milk or whey by cation exchange chromatography[41] which has been scaled up for industrial application[42]. The method isolates LPO together with lactotransferrin (Lf) which is also cationic at neutral pH. LPO and Lf can also be separated by carboxymethyl cation exchange chromatography on CM-Toyopearl[43]. LPO has been isolated from whey by gel filtration and hydrophobic chromatography on Butyl Toyopearl 650 M[44].

Lactotransferrin

The transferrins are a group of specific metal-binding proteins, the best characterized of which are: serotransferrin, ovotransferrin and lactotransferrin.

Human colostrum and milk contain 6-8 mg/ml and 2-4 mg/ml lactotransferrin, respectively, representing ~ 25% of the total protein in the latter; bovine colostrum and milk contain ~ 1 and 0.02-0.35 mg/ml, respectively[45]. Because the concentration of Lf in human milk is considerably higher than that in bovine milk, there is considerable interest in supplementing bovine milk-based infant formulae with bovine Lf. Bovine lactotransferrin has also been considered for use in food as an antiseptic or bacteriostatic, in feed as the growth promoter and in medicine as a chemical mediator or iron supplier.

Lfs have been isolated from the milks of several species

and some of the isolation procedures have industrial scale potential. As stated above, under appropriate conditions cation exchange resins simultaneously bind Lf and LPO and procedures have been developed for their separate elution. Other methods reported in the literature for the isolation of lactoferrin include gel filtration[46] and various affinity chromatographic procedures using immobilized heparin[47], ferritin[48], triazine dyes[49] and monoclonal antibodies[50].

Immunoglobulins

Immunoglobulins (Igs), one of the principal defence methanisms of the body, are present in the mammary secretions, especially colostrum, of all mammalian species. Bovine colostrum protein contains ~10% Ig but this level decreases to ~0.1% within about a week post-partum.

In situations where it is not possible to feed colostrum to neonatal ruminants and pigs, an alternative source of Ig is necessary and therefore there is interest in the production of Ig concentrates for this purpose. Calf milk replacers and piglet and lamb supplements enriched with Ig are available.

The classical method for preparing Ig is by salting out, usually with $(NH_4)_2SO_4$. This method is effective but expensive and current commercial products are usually prepared by ultrafiltration of colostrum or milk from hyperimmunized cows[51-53]. Other methods for the isolation of Ig, sometimes with Lf, use ultrafiltration in combination with ion exchange chromatography[54,55], immobilized monoclonal antibodies[56], metal chelate[57] or gel filtration[46] chromatography.

Although human infants are not able to absorb Ig from the intestine, Ig still plays an important defensive role in reducing the incidence of intestinal infection. There is general agreement on the superiority of breast feeding for healthy full-term infants but it is frequently impossible to breast-feed preterm or very-low-birth-weight infants, who may be fed on banked human milk. However, such infants have high protein and energy requirements which may not be met by human milk and consequently special formulae have been developed. A "milk immunological concentrate", prepared by diafiltration of acid whey from colostrum and early lactation milk from immunized cows, for use in such formulae has been described[58]. The final product contained ~75% protein, 50% of which was Ig, mainly IgG_1 and not IgA, which is predominant in human milk.

Glycomacropeptide

Anion exchange chromatographic[59,60] and other processes[61] are also being developed for the recovery of the glycomacropeptide split-off from casein by rennet type enzymes and present in sweet whey at a concentration of ~1.2-1.5 mg/ml. The interest in this peptide stems from its unusual amino acid profile, e.g. it contains no Phe, Tyr, Trp, Lys or ½Cys; the absence of aromatic amino acids makes it very suitable ror the nutrition of patients suffering from phenylketonuria.

CO-PRECIPITATE PRODUCTION

The methods described above are used to recover casein and whey protein products and fractions separately. However, caseins and whey proteins can be co-precipitated by first heating milk, at its natural pH, to temperatures that denature the whey proteins and induce their complexation with casein, followed by precipitation of the milk protein complex by acidification to pH 4.6 or by a combination of added $CaCl_2$ and acidification[62-64]. Products produced in this manner are referred to as casein-whey protein co-precipitates. Yields of 92-98% of total milk protein are obtained, compared to < 80% for acid or rennet caseins[65]. Co-precipitates produced by these methods have poor solubility properties[66]; however, processes for the manufacture of similar products with good solubility have been developed[67-69]. These processes involve preadjusting milk to pH 7.0-7.5 heating at 90°C x 15 min or preadjusting milk to pH 10 and heating at 60°C x 3 min prior to isoelectric precipitation.

Again the isoelectric or pH/$CaCl_2$ induced co-precipitates are processed in a similar manner to that used for caseins and these products may also be converted to proteinates (usually the sodium salts) by addition of base.

PRODUCTION OF MILK PROTEIN CONCENTRATES

Skim milk may also be processed directly by ultrafiltration/diafiltration to yield milk protein concentrates (MPC) that contain a range of protein contents up to ~ 80% and in which the casein is in a similar micellar form to that found in milk while the whey proteins are also reported to be in their native form[70,71]. The products have a relatively high ash content, since protein-bound minerals are retained.

CHEMICALLY, PHYSICALLY AND ENZYMATICALLY MODIFIED MILK PROTEINS

The acceptability of chemically-modified proteins for use in foods is doubtful, at least at present, but enzymatically or physically modified proteins are acceptable. Many possibilities exist for improving the functional properties of milk proteins by enzymatic and physical modification methods.

Physical Modification of Milk Proteins

At least one commercial processing operation intentionally uses physical modification to produce a milk protein-based food ingredient. In this process, a proteinaceous, water dispersible macrocolloid, comprising of substantially non-aggregated particles of whey protein, that has a smooth emulsion-like organoleptic character when hydrated, is produced[72,73]. The process involves simultaneously heating to temperatures of ≥ 80°C and extensively shearing a whey protein concentrate solution, containing about 40-50% (w/w) total solids of which about 45 to 55% is undenatured protein and which is adjusted to a pH in the range 3.7 to 4.2.

During the simultaneous heating and shearing, denaturation of the whey proteins occurs but the shearing action prevents the formation of substantial amounts of aggregated, denatured protein particles larger than about 2.0 μm. Therefore, the substantially spherical particles formed have a mean diameter particle size distribution in a dried state ranging from about 0.1 to 2 μM with less than about 2% of the total number of particles exceeding 3 μm in diameter. The micro-particulated protein product thus produced is marketed under the brand name Simpless^R by the NutraSweet Company and due to its creamy sensory properties it is proposed as a fat substitute in various foods. A number of other whey protein based fat substitutes have been described[74,75].

Milk Protein Hydrolyzates

There is continuing interest in the development of milk protein hydrolyzates tailored for use in specific applications in the health-care, pharmaceutical, baby food and consumer product areas. Several methods have been described in the literature for the production, characterization and evaluation of milk protein hydrolyzates.

Peptides with various types of biological activity have been isolated from the hydrolyzates of casein and whey proteins[76,77]. At least some of these peptides are produced in vivo and therefore may play a physiological role in vivo. There might be commercial interest in producing these peptides from purified milk proteins on an industrial scale in the future.

GENETICALLY ENGINEERED MILK PROTEINS

Reports on the genetic engineering of casein and whey proteins[78-82] are now appearing in the literature and in the future genetic engineering may be a very important method by which to tailor functionality and further expand the range of functional milk protein products.

Bibliography

1. P.F. Fox, 1989, in, "Developments in Dairy Chemistry - 4 - Functional Milk Proteins", P.F. Fox, ed., Elsevier Applied Science, London, p. 1.

2. M.M. Hewedi, D.M. Mulvihill and P.F. Fox, Ir. J. Food Sci. Technol., 1985, 9, 11.

3. J.E. Hoff, S.S. Nielsen, I.C. Peng and J.V. Chambers, J. Dairy Sci., 1987, 70, 1785.

4. D.A. Lonergan, J. Food Sci., 1983, 48, 1817.

5. D.A. Lonergan, 1984, US Patent 4 462, 932.

6. F. Morel, Process, 1991, No. 1059, 53.

7. R. Noel, 1991, French Patent Application, FR 2 657 233 A1.

8. J.L. Maubois and G. Ollivier, International Dairy Federation, Brussels, 1991, B-Document 213, p 7.

9. D.M. Mulvihill, in, "Developments in Dairy Chemistry - 4 - Functional Milk Proteins", P.F. Fox, ed., Elsevier Applied Science, London, 1989, p 97.

10. E.M. Allen, A.G. McAuliffe and W.J. Donnelly, Ir. J. Food Sci. Technol., 1985, 9, 85.

11. E. Terre, J.L. Maubois, G. Brule and A. Pierre, French Patent, 1986, FR 2 592 769.

12. M.H. Famelart, C. Hardy and G. Brule, Le Lait, 1986, 69, 47.

13. J.M. Murphy and P.F. Fox, Food Chemistry, 1990, 39, 27.

14. C.V. Morr, in, "Developments in Dairy Chemistry - 4 - Functional Milk Proteins", P.F. Fox, ed., Elsevier Applied Science, London, 1989, p 245.

15. R.R. Zall, A. Kuipers, L.L. Muller and K.R. Marshall, N.Z. J. Dairy Sci. Technol., 1979, 14, 79.

16. K.R. Marshall, in, "Developments in Dairy Chemistry - 1 - Proteins", P.F. Fox, ed., Applied Science, London, 1982, p 339.

17. J.F. Hayes, J.A. Durkerley, L.L. Muller and A.T. Griffin, Aust. J. Dairy Technol., 1974, 29, 132.

18. B.R. Breslau, R.A. Cross and J. Gaulet, J. Dairy Sci., 1975, 58, 782.

19. D.N. Lee and R.L. Merson, J. Food Sci., 1976, 41, 778.

20. J.N. de Wit, G. Klarenbeek and R. de Boer, Proceedings of International Dairy Congress, Paris, Congruilait, Paris, 1978, p 919.

21. M.E. Matthews, R.K. Doughty and J.L. Short, N.Z. J. Dairy Sci. Technol., 1978, 13, 216.

22. L.L. Muller and W.J. Harper, J. Agric. Food Chem., 1979, 27, 662.

23. J.L. Maubois, A. Pierre, J. Fauquant and M. Piat, International Dairy Federation, Brussels, Bulletin 212, 1987, p 154.

24. J. Patoka and P. Jelen, J. Food Sci., 1987, 52, 1241.

25. S.H. Kim, C.V. Morr and J.G. Surak, J. Food Sci., 1989, 54 25.

26. K.J. Burgess and J. Kelly, J. Food Technol., 1979, 14, 325.

27. D.E. Palmer, in, "Food Proteins", P.F. Fox and J.J. Condon, eds., Applied Science, London, 1982, p 341.

28. B. Mirabel, Annales de la Nutrition et de l'Alimentation, 1978, 23, 243.

29. J. Kaczmarek, in, "Proceedings 1980 Whey Production Conference", ADPI, Chicago, USDA Philadelphia, PA, 1980, p 68.

30. B.P. Robinson, J.L. Short and K.R. Marshall, N.Z. J. Dairy Sci. Technol., 1976, 11, 114.

31. R.J. Pearce, Aust. J. Dairy Technol., 1983, 38, 144.

32. R.J. Pearse, International Dairy Federation, Brussels, Bulletin 212, 1987, p 150.

33. A. Pierre and J. Fauquant, Le Lait, 1986, 66, 405.

34. C.H. Amundson, S. Watanawanichakorn and C.G. Hill, J. Food Proc. Preserv., 1982, 6, 55.

35. T. Kuwata, A.M. Pham, C.Y. Ma and S. Nakai, J. Food Sci., 1985, 50, 605.

36. S.A. Al-Mashikhi and S. Nakai, J. Food Sci., 1987, 52, 1237.

37. P. Mailliart and B. Ribadeau-Dumas, J. Food Sci., 1988, 53, 743.

38. K.K. Fox, V.H. Holsinger, L.P. Posati and M.J. Pallansch, J. Dairy Sci., 1967, 50, 1363.

39. J.N. de Wit, G. Klarenbeek and M. Adamse, Neth. Milk Dairy. J., 1986, 40, 41.

40. R. Hayashi, Y. Kawamura and S. Kunugi, J. Food Sci., 1987, 52, 1107.

41. K.G. Paul, P.I. Ohlsson and A. Hendriksson, FEBS Lett, 1980, 110, 200.

42. J.P. Prieels and R. Peiffer, UK Patent Application, 1986, GB2, 171, 102, A1.

43. S. Yoshida and Ye-Xiuyun, J. Dairy Sci., 1991, 74, 1439.

44. S. Yoshida, J. Dairy Sci., 1988, 71, 2021.

45. B. Reiter, in, "Developments in Dairy Chemistry - 3 - Lactose and Minor Constituents", P.F. Fox, ed., Elsevier Applied Science, London, 1985, p 281.

46. S.A. Al-Mashiki and S. Nakai, J. Dairy Sci., 1987, 70, 2486.

47. L. Blckberg and O. Hernell, Fed. Eur. Biol. Soc. Lett., 1980, 109, 180.

48. J.J. Pahud and H. Hilpert, Protides Biol. Fluids, 1976, 23, 571.

49. K. Shimazaki and N. Nishio, J. Dairy Sci., 1991, 74, 404.

50. H. Kawakami, H. Shimmoto, S. Dosoko and Y. Sogo, J. Dairy Sci., 1987, 70, 752.

51. N. Kothe, H. Dichtelmuller, W. Stephan and B. Eichentopf, European Patent, 1986, 0 173 999 A2.

52. G.H. Scott and D.O. Lucas, European Patent, 1987, 0 239 722 A1.

53. H. Tamguchi, M. Goto, T. Sakuchi, T. Ano, O. Kirchara and K. Ando, European Patent Application, 1990, 0 391 416 A1.

54. E. Dubois, French Patent, 1986, FR 2 605 322.

55. R.C. Bottomley, European Patent Application, 1989, 0 320 152 A2.

56. M.M. Gani, K. May and K. Porter, European Patent, 1982, 0 059 598 A1.

57. S.A. Al-Mashiki, E. Li-Chan and S. Nakai, J. Dairy Sci., 1988, 71, 1747.

58. H. Hilpert, in, "Human Milk Banking", A.F. Williams and J.B. Baum, eds., Raven Press, New York, 1984, p 17.

59. P.J. Skudder, J. Dairy Res., 1985, 52, 167.

60. J. Burton and P.J. Skudder, UK Patent Application, 1987, 21 88 526A.

61. S.C. Marshall, CSIRO Food Res. Quart, 1991, 51, 86.

62. R.A. Buchanan, N.S. Snow and J.F. Hayes, Aust. J. Dairy Technol., 1965, 20, 139.

63. L.L. Muller, N.S. Snow, J.F. Hayes and R.A. Buchanan, Proc. XVII Intern. Dairy Congr., Munich, 1966, Vol. E/F, p 69.

64. L.L. Muller, J.F. Hayes and N.S. Snow, Aust. J. Dairy Technol., 1967, 22, 12.

65. C.R. Southward and R.M. Aird, N.Z. J. Dairy Sci. Technol., 1978, 13, 77.

66. C. Towler, N.Z. J. Dairy Sci. Technol., 1974, 9, 155.

67. P.B. Connolly, International Patent Application, 1982, PCT/US, 81/01357.

68. J.M.G. Lankveldt, in, "Milk Proteins, '84", Proc. Intern Congr. on Milk Proteins, Luxembourg, T.E. Galesloot and B.J. Tindbergen, eds., Pudoc Publishers, Wageningen, 1984, p 269.

69. M.B. Grufferty and D.M. Mulvihill, J. Soc. Dairy Technol., 1987, 40, 82.

70. A. Novak, International Dairy Federation, Brussels, B. Doc. 213, 1991, p 32.

71. Z. Puhan, in, "Proceedings of Dairy Products Technical Conference", Chicago, ADPI Chicago, 1990, p 33.

72. N.S. Singer, S. Yamamoto and J. Latella, U.S. Patent, 1988, 4 734 287.

73. N.S. Singer and J.M. Dunne, J. Am. College Nutr., 1990, 9, 388.

74. F.A. Groves and C.M. Niemand, European Patent, 1987, O 129 346.

75. W. Rattray, M.Sc. Thesis, 1992, National University of Ireland.

76. H. Meisel, H. Frister and E. Schlimme, Z. Ernahrungswiss, 1989, 28, 267.

77. J.L. Maubois and J. Leonil, Le Lait, 1989, 69, 245.

78. R. Jimenez-Flores and T. Richardson, J. Dairy Sci., 1988, 71, 2460.

79. R.A. McKnight, R. Jimenez-Flores, Y.C. Kang, L.K. Creamer and T. Richardson, J. Dairy Sci., 1989, <u>72</u>, 2464.

80. T. Richardson, S. Oh, R. Jimenez-Flores, T.F. Kumosenski, E.M. Brown and H.M. Farrell, Jr., in "Advanced Dairy Chemistry - 1 - Proteins", P.F. Fox, ed., Elsevier Applied Science, London, 1992, p 545.

81. C.A. Batt, L.D. Robson, D.W. Wong and J.E. Kinsella, Agric. Biol. Chem., 1990, <u>54</u>, 949.

82. S. Lee, Y. Cho and C.A. Batt, J. Agric. Food Chem., 1993, <u>41</u>, 1343.

Protein Engineering Studies of ß-Lactoglobulin

L. Sawyer[1], J. H. Morais Cabral[1], and C. A. Batt[2]

[1] THE EDINBURGH CENTRE FOR MOLECULAR RECOGNITION AND DEPARTMENT OF BIOCHEMISTRY, THE UNIVERSITY OF EDINBURGH, EDINBURGH EH8 9XD, UK

[2] DEPARTMENT OF FOOD SCIENCE, CORNELL UNIVERSITY, ITHACA, NY 14853, USA

1 INTRODUCTION

The purpose of this article is to review some of the site-directed mutagenesis work which has been performed to investigate the basic properties of BLG, the putative function and finally the processing behaviour of this small milk-whey protein.

2 BACKGROUND

For many years the abundant milk protein β-lactoglobulin (BLG) has been the subject of detailed study in solution where it has served as a convenient test system for a wide variety of techniques.[1] Despite all of this work no function has been definitively ascribed to it although several have been suggested.[2] However, when the structure of one crystal form was determined in Edinburgh and shown to be similar to plasma retinol-binding protein (RBP),[3,4] this prompted the speculation that the function of BLG was as a transport protein, possibly for retinol,[5-7] specific receptors for BLG-retinol having been found in neonate calf intestine.[3] The suggestion that the function of BLG may be one of transporting hydrophobic molecules gains further support from the structures of insecticyanin[8] and of bilin-binding protein[9] which bear close 3-dimensional similarities to BLG and RBP.[10] Indeed, this particular fold appears to occur much more widely than was once thought. For instance, Pervais & Brew[6] first compared the sequences of BLG with those of α-1-microglobulin (protein HC) and α-2u-globulin (also, mouse major urinary protein,

MUP[11,12]), and the family continues to grow. (Figure 1).

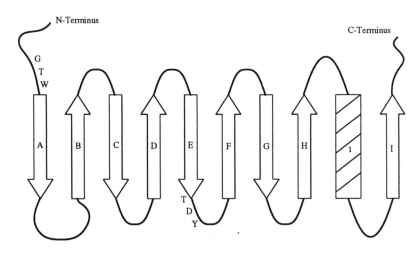

Figure 1 Schematic diagram of the antiparallel β-sheet structure of the lipocalin fold. The arrows represent the β-strands, the hatched box is the α-helix and the positions of two of the characteristic sequence motifs are shown at the start of strand A and the end of strand E.

Now, the lipocalin family contains about 20 small (160-180 residues) proteins, most of which are found in secretions like milk, urine and mucous.[13-15] The majority of the lipocalins are able to bind small, hydrophobic molecules in their internal binding pocket and some form of transport or transduction function is suspected. Only one member has so far been found which has enzymic activity, although it is possible that another, apolipoprotein D, has proteolytic activity.[16] The smaller, intracellular, binding proteins like fatty acid binding protein and p2-myelin also comprise a β-barrel, in this case with 10 strands, and because of some indications of sequence similarity in the N-terminal region, may be more distant relatives forming the calycin superfamily.[15]

BLG binds retinol somewhat more tightly than does RBP [17,18] and model-building studies with BLG and retinol show not only that the hydrophobic binding pocket can readily accommodate retinol, but also that there may be more space at the inner end to allow somewhat larger molecules to be bound and protected from the external solvent. There are several reports of the binding of hydrophobic molecules to BLG [17,19-21] including SDS. Intriguingly, crystal studies on another crystal form appear to indicate an alternative retinol-binding site to that expected[22] and unpublished observations from work in Edinburgh on this same crystal form soaked in butyl iodide also show this unexpected binding. Binding of p-nitrophenylphosphate[21] gives an indication of yet another binding site.[1] However, it is clear that BLG binds hydrophobic molecules with quite high affinity, typical values for K_d being around 10^{-4}-10^{-6}M.

BLG is a remarkably acid-stable protein. Aschaffenburg & Drewry[23] used this fact to prepare the protein by lowering the pH of whey to 2 at which point effectively all other proteins are denatured. Further, BLG survives the enzymic digestion in the stomach to appear intact in the intestine.[24-26] BLG is known to pass through the stomach unscathed and is thus a good candidate for investigation as a carrier of small, labile, and/or toxic substances at least as far as the intestine.

A variety of other properties have been identified relating to the behaviour in solution. For example, there are 5 Cys residues of which 4 form disulphide bridges between 66-160 and 106-119. The free sulphydryl group at 121 has been the subject of some controversy associated with the possibility of disulphide interchange between 119 and 121.[27] Titration work by Tanford[28] has revealed a carboxyl group with a pK value typical of histidine and, furthermore, a series of distinct conformational changes exist between pH 3 and 10, as monitored by optical rotatory dispersion[29] and sulphydryl reactivity.[30]

That some of these properties have a direct bearing on the behaviour of BLG in solution at physiological temperatures, but more particularly during heating, is well known to the food technologist.[31] As the dairy industry seeks to increase yields, of cheese for example, by incorporating whey proteins, the problem of the irreversible denaturation and aggregation of BLG needs to be addressed since this not only affects the heat treatment but also the proteolysis and gelation properties. Calorimetric studies on the thermal denaturation of BLG over a wide range of pH and, more recently, concentration, show that the denaturation process at the conditions prevailing in milk, pH 6.5 and 3 mg ml^{-1}, is most complex.[32-34] As a result of such calorimetric measurements on the denaturation of BLG, the approach to whey protein incorporation based upon forwarming temperatures up to 90°C have been abandoned since denaturation occurs at around 70°C.

One obvious direction in which to proceed, therefore, involves not a change in the processing protocols but rather a change in the milk components themselves. Protein engineering can certainly offer considerable benefits to the dairy industry, although, it must be admitted, there are significant problems associated both with the generation and the public acceptance of transgenic animals.

Recent advances in molecular biological manipulation have allowed specific mutations to be introduced into a protein's amino acid sequence in such a way as to mimic, and to some extent, speed up natural selection. More importantly, mutations which would not normally provide selective advantage in nature can be introduced in a directed and deliberate manner. Several types of mutation can be considered: those which are introduced to try to modify a function,[35] those which are introduced to modify the stability of a protein[36] and those which are designed to probe a catalytic mechanism or a binding function.[37] In almost every case, unlike the naturally occurring mutations where the odd millenium matters little, successful site-directed modification has required knowledge of the 3-dimensional structure of the protein, as well as convenient protocols for overexpressing, purifying and characterising the mutant proteins. These prerequisites are adequately met with BLG.

3 THE STRUCTURE OF β-LACTOGLOBULIN

We have determined the crystal structure of two forms, Y (orthorhombic prisms, spacegroup $P2_12_12_1$, a=55.7, b=67.2, c=81.7Å, $\alpha=\beta=\gamma=90°$)[3] and a re-interpretation of Z (trigonal blocks, spacegroup $P3_221$, a=b=54.2, c=113.4Å, $\gamma=120°$)[22] of BLG at pH 7.8 to 2.8Å and 3.5Å resolution respectively, and are at present extending these studies to higher resolution. We have found that the resolution of the data obtainable in the laboratory is limited, but that using a synchrotron, higher resolution data, especially of the Y form, can be measured. However, there is a considerable amount of thermal diffuse scattering[38] which indicates that the molecules in the crystal are in significantly greater motion than might be expected. It is not yet clear whether this also reflects intramolecular motion, possibly associated with conformational changes. Data have been collected on the Y form to 1.8Å resolution at the synchrotron at the SERC Daresbury Laboratory and processed to give a set which is >90% complete with an $R_{merge} = 0.09$. Refinement using these data alone or in conjunction with lower resolution data has not gone smoothly and we have recollected both the low-to-medium resolution shell (2.8Å resolution on lattice Y collected on an area detector), have reprocessed all of the heavy atom data and recalculated a map at a nominal 2.8Å resolution. This is currently being reinterpreted in the light of the recent results on the Z form.

The structure of the lattice Z form of the BLG was published several years ago[22] following that of the Y form.[3] Whilst both forms indicated a conserved core structure of β-sheet forming a conical barrel or calyx, the difficulty in obtaining convincing refined coordinates of the Y form, despite the high resolution data, indicated that some details were incorrect. Whilst recollecting and reprocessing much of the Y-form data, we have carried out a medium resolution structure redetermination of the Z form which has highlighted the problem in both high pH forms. The basic scheme of the β-structure for the protein is shown in Figure 2 and remains unchanged, but the threading of the polypeptide chain through the molecule has been altered between residues 30 and 60 whereby a large and poorly defined loop has become smaller and the extra 5 residues removed at this point are drawn back through the strands to increase the size of the loop around residue 30. This has a dramatic effect upon the distribution of the residues, and makes the hydrophobicity profile of the molecule derived by the algorithm of Eisenberg et al.[39] change from one which shows a severe incompatibility between the residues and their expected environments, to one which is highly satisfactory. Fortunately, none of these changes affects the results or their interpretation of the mutations which are discussed below.

Because of the large number of distinct crystal forms of BLG which exist, the various conformational changes observed as the pH is raised between 3 and 8 ought to be observable in molecular detail. There are good data available for the structure at pH 6.5 of the X form, and the structure at this pH is currently being tackled in Leeds by A.C.T.North and collaborators, but the crystals at lower pH seem to have a poor crystallinity and this in turn reduces the resolution of the X-ray diffraction data obtainable. For example, at pH 3, crystals (hexagonal

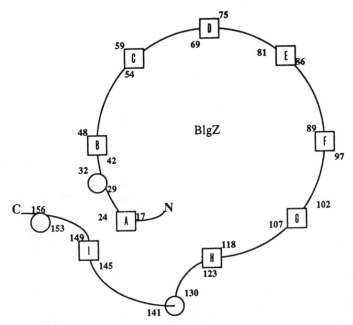

Figure 2 A cartoon showing the assignment of the β-sheet in the current, medium resolution structure of β-lactoglobulin, viewed 'into' the calyx. The exact lengths of the various stretches of secondary structure may still alter as the structure is refined.

prisms, space group P6$_3$, $a=b=68.5$, $c=143.2$Å, $\gamma=120°$) have been found to diffract only to a modest resolution between 3-3.5Å and, unlike the higher pH forms, the resolution does not improve with synchrotron radiation. The crystal structure of the pH 3 form of BLG, for which we have recently collected X-ray data, will allow us to compare it with the high pH, protease labile forms and perhaps gain some insight into the reasons for the remarkable acid stability.

4 CLONING AND EXPRESSION

Amongst the several reports of the cloning and expression of BLG, two report its significant overexpression. That from Cornell used E.coli to overexpress bovine BLG with an additional N-terminal methionine using the tac promoter vector pTTQ18.[40-42] Some 15% of the cell protein was BLG but most of this was produced as inclusion bodies which required solubilisation in guanidine hydrochloride with subsequent renaturation. However, the BLG so obtained is indistinguishable from the protein obtained from milk. Totsuka et al.[43] used a different approach by cloning the coding sequence for pre-BLG between the promoter and terminator regions of the yeast glyceraldehyde-3-phosphate dehydrogenase gene of pYG100. BLG was expressed and secreted in its natural form, the signal sequence having been removed, at a rate of around 1-2 mg l^{-1} of S.cerevisiae culture supernatant. In Edinburgh, we have also used a yeast expression system, this time under control of the phosphoglycerate kinase promoter pMA91, to produce native ovine BLG at around 4-6mg l^{-1} S.cerevisiae culture supernatant. Mutants of this last protein are generated in M13 and subcloned into the expression vector, pMABLG.[44,45]

5 BINDING STUDIES

The binding of retinol to BLG may or may not be physiologically important, but it might be expected to bind to the protein in the same way as it does to RBP, ie. in the central cavity. The only direct published evidence,[22] however, indicates a binding site along the cleft formed by the helix and the outer surface of the β-barrel. Implicated in the interaction with retinol, however, is a lysine residue: Lys70 in the modelled BLG-retinol complex,[3] Lys141 in the putative X-ray model. Mutants K70M and K141M were produced and only the former exhibited a marked decrease in binding as observed by protein fluorescence. Further, K70M lacked the characteristic blue shifted spectrum for the protonated Schiff base retinylidenepropylamine, which was observed for K141M and the wild-type BLGs. Final proof was given by mass spectrometry which revealed the presence of the retinol covalently bound to a tryptic peptide which contained the Lys70 residue. Clear separation from the peptide which contained Lys141 was achieved, providing an unambiguous assignment of the retinol to the internal binding pocket, as expected.[4]

Because of the acid and protease stability of BLG, a possible drug delivery role can be envisaged for the protein, although more convenient alternatives are already available. Tailoring a binding pocket to carry small, toxic, insoluble or labile molecules is an exercise of considerable interest scientifically, if not commercially. The first stage of such a study is to model the small molecule into the protein and identify which residues need to be altered and to what. The creation of a different disposition of side chains may lead to problems of incorrect folding, insolubility or instability and these have yet to be overcome in the case of BLG.[26,45]

6 THE SULPHYDRYL GROUP

The events leading to the macromolecular association observed with BLG appear to be mediated by sulphide-disulfide interactions. A large number of indirect experimental results, using a variety of reagents which modify free thiols, point to a reduction in aggregation and/or gelation of the protein. Thus, blocking the free thiol at Cys121 can prevent gelation and therefore stabilize BLG against thermally induced alterations. We have obtained evidence both from natural BLG and genetically modified forms of the protein to support these observations. Porcine BLG which does not contain a free thiol does form aggregates upon heating.[46] Further, porcine BLG does not induce, or participate in, the polymerization of bovine BLG, suggesting that macromolecular aggregation is not a chain-reaction capable of being self-sustaining without a free thiol group on the protein. The mutants C119S and C121S were made in Edinburgh and whilst the former was expressed, it was not secreted from the yeast cells, whilst no such problem was apparently encountered with the latter. This is in agreement with the X-ray work where no evidence of either the alternative or a mixture of disulphides is obtained.[3,22,45] In similar experiments in Cornell, a recombinant form of BLG was engineered to eliminate the free thiol group. Whilst the original intent was to create a variant without the Cys121, in practice this variant protein could not be refolded and purified. A C121A variant constructed by oligonucleotide mediated mutagenesis was produced but subsequent denaturation and renaturation failed to yield a properly refolded protein. Aggregates were consistently obtained under conditions where all other BLG mutants yielded fully folded, acid-stable protein. This is an interesting difference between the yeast and E.coli expression systems.

As an alternative, we chose to form the free thiol into a third disulphide bond by the substitution of another amino acid with cysteine. Two mutations were designed in a rational manner on the basis of computer graphics observations and then subjected to molecular dynamics simulation to test the predicted hierarchy of their thermal stability. Both Leu104Cys and Ala132Cys substitutions placed thiols close enough to Cys121 to permit the formation of a new disulphide bond. However, the simulated molecular dynamic behaviour of the two proposed mutant forms of BLG differed both with respect to fluctuation in the positions of the backbone atoms and also the molecular radius of gyration. By both criteria, the L104C mutation appeared to be the closer to the wild-type structure.

Each mutant protein was produced in sufficient quantity to test its thermostability in the recombinant E.coli system at Cornell and each was purified from inclusion bodies. The purification procedure employs a denaturation step followed by a renaturation in the presence of both reduced and oxidized glutathione. The refolded BLG was further purified on the basis of its acid stability and finally by gel filtration. Yields typically were only 10 mg l^{-1} of culture but this was sufficient for all of the requisite analyses. A number of indirect measures of protein structure including circular dichroism and intrinsic fluorescence revealed that the recombinant BLG was indistinguishable from the protein isolated from bovine milk. We have produced small crystals of the recombinant protein which are not yet suitable for X-ray analysis, but which appear similar in form to the protein prepared from milk.

A key concern was to verify the formation of the correct disulfide linkages. Although examination of the model and computational analysis predicted the ability of the newly introduced cysteine residues to link with the Cys121, formal proof of their existence was necessary. Formation of disulfide linkages in the correct arrangement, as designed, was confirmed by peptide mapping. At the resolution which could be acheived by HPLC peptide mapping, a third disulfide was identified which consisted of the Cys121 joined to the newly introduced Cys either at 104 (L104C) or at 132 (A132C). In contrast to wild-type BLG, which polymerizes at temperatures >65°C, SDS-PAGE showed that neither of the mutant proteins polymerized. The conformational stability of the L104C and A132C mutant proteins against thermal denaturation had also been substantially increased (by some 8-10°C) compared with the wild-type protein. Furthermore, the A132C BLG exhibited an enhanced stability against denaturation by guanidine hydrochloride compared to that of both wild-type BLG and L104C BLG.[47]

Demonstrating the ability to reduce thermal aggregation clearly pointed the way toward improving gelation. Gelation is the controlled aggregation of monomers resulting in the formation of a complex network that entraps water. Although BLG can form thermoset gels, it does so only at concentrations >10% and at temperatures above 85°C, depending upon the conditions of ionic strength, pH, protein concentration and type of cation. This limits the usefulness (and hence the value) of BLG as a food ingredient where the objective is to increase the viscosity of a solution. The gelation characteristics of bovine BLG A have been enhanced by the selective introduction of cysteine substitutions to increase the free thiol content of the protein.[48] A recombinant version of bovine BLG A has been modified by creating Arg40Cys, Phe82Cys, and the double R40C/F82C variants. Titration showed that, as expected, the number of free thiols increased correspondingly, suggesting that additional disulfide linkages are not formed. The strength of gels formed by heating at 70-90°C was measured using a micro-scale penetration test.[49] F82C and R40C/F82C displayed a gel strength equivalent to wild-type BLG at a much lower concentration compared to wild-type BLG. R40C could not be brought to a sufficient concentration (>5%) without the formation of insoluble aggregates. Increasing the free thiol content also enhanced the ability of BLG to form high molecular weight aggregates as observed during the heating of milk. An unexpected result, however, was that the introduction of an extra free thiol also increased the ability of these BLG variants to be cleaved by chymosin. Further, these variants were more susceptible to acid precipitation. These latter observations may be important in improving the performance of BLG during the renneting process. In general, the underlying cause of these differences in stability will be found in the subtle modifications to the overall structural integrity of the mutant proteins when compared to the wild-type (native) version.

The increased sensitivity of the double mutant BLG (R40C/F82C) to acid, in conjunction with its enhanced gelation properties, suggested that this protein might serve as a useful ingredient in dairy products where the formation of a gel is critical to product quality. Yogurt was selected as a model system because not only does it represent a relatively simple dairy product but also the formation of a thermoset gel is critical for this product, and failure of the gel to form is considered a defect. A yogurt product was formulated using milk fortified with skim milk powder and R40C/F82C mutant BLG was added to a final concentration of 0.075%. These solutions were heated either at 85°C or 70°C, the former being the normal processing temperature. The heated solutions were then fermented using a combination of Lactobacillus bulgaricus and Streptococcus thermophilus. Yogurts containing R40C/F82C mutant BLG exhibited less whey syneresis in formulations processed at 70°C when compared to yogurts to which wild-type BLG was added.[50] This reduced whey syneresis resulted from the formation of a much stronger gel which was capable of retaining more water. Extensive studies have been carried out to improve the functional properties of whey-protein concentrates used for yogurt manufacture; however there is still a continuing need to formulate novel ingredients. The introduction of additional free thiols to BLG appears to improve its functionality as an ingredient for yogurt manufacture and to reduce processing costs.

8 CONCLUSIONS

The annual literature on BLG is extensive and covers too wide a range of topics, few of which have so far involved the mutagenesis of the protein, for this review. Many of these studies involve the behaviour of BLG alone, in simple binary mixtures or in the more complex mixtures found in milk, but under conditions similar to those found in food processing. The studies above show that site-directed modification of BLG can have both predictable and useful consequences on the processing behaviour and consequently additional efforts to probe the relationship between the structural elements of BLG and its functional properties should prove fruitful.

9 ACKNOWLEDGEMENTS

The authors are grateful to John Brady, Yunje Cho, Sam-Pin Lee, Steve Watkins, Wei Gu, Linda Gilmore, Gary Paterson, Thales Rocha, Alan McAlpine, Tony North and Carl Holt for help with and discussions of the work described here, which was supported by the Science Engineering Research Council, the European Commission Biotechnology Programme, the National Dairy Promotion and Research Board, the New York State Science and Technology Foundation, the US Army Research Office and the National Science Foundation. This work has been facilitated by a NATO Travel Award.

10 REFERENCES

1. S.G. Hambling, A.S.McAlpine and L.Sawyer 'Advanced Dairy Chemistry I' ed. P.F.Fox, Elsevier. London, 1992, Chapter 4, p.140
2. M.P. Thompson and H.M. Farrell,Jnr, 'Lactation. A Comprehensive Treatise', 1974, Vol III,p.109
3. M.Z. Papiz, L. Sawyer, E.E. Eliopoulos, A.C.T. North, J.B.C. Findlay, R. Sivaprasadarao, T.A. Jones, M.E. Newcomer and P.J. Kraulis, Nature ,1986, 324, 383
4. M.E. Newcomer, T.A. Jones, J. Aqvist, J. Sundelin, U. Erikkson, L. Rask and P.A. Peterson, EMBO J., 1984, 3, 1451
5. L. Sawyer, M.Z. Papiz, E.E. Eliopoulos and A.C.T. North, Biochem.Soc.Trans., 1985, 13, 265
6. S. Pervaiz and K. Brew, Science, 1985, 228, 335
7. J. Godovac-Zimmermann, A. Conti, J. Liberatori and G. Braunitzer, Biol.Chem.Hoppe-Seyler, 1985, 366, 431
8. H.M. Holden, W.R. Rypniewski, J.H. Law, and I. Rayment, EMBO J., 1987, 6, 1565
9. R. Huber, M. Schneider, I. Mayer, R. Muller, R. Deutzmann, F. Suter, H. Zuber, H. Falk and H. Kayser, J.Mol.Biol., 1987, 198, 499
10. L. Sawyer, Nature, 1987, 327, 659
11. A.J.Clark, P.M. Clissold, A. Shawi, P. Beattie and J.O. Bishop, EMBO J., 1984, 3, 20
12. Z. Böcskei, C.R. Groom, D.R. Flower, C.E.Wright, S.E.V. Phillips, A. Cavaggioni, J.B.C. Findlay and A.C.T. North, Nature, 1992, 360, 186.
13. J. Godovac-Zimmermann, TIBS, 1988, 13, 64.
14. A.C.T. North, Biochem.Soc.Symp., 1991, 57, 35
15. D.R. Flower, A.C.T. North and T.K. Atwood, Prot.Sci, 1993, 2, 753
16. L. Kesner, W. Yu and H.L. Bradlow, Ann.N.Y.Acad.Sci., 1990, 586, 198
17. S. Futterman and J. Heller, J.Biol.Chem., 1972, 247,168
18. R.D. Fugate and P.S. Song, Biochim.Biophys.Acta, 1980, 625, 28
19. T.S. Seibles, Biochemistry, 1969, 8, 2949
20. W.L. Stone and A. Wishnia, Bioinorg.Chem., 1978, 8, 517
21. H.M. Farrell,Jr, M.J. Behe and J.A. Enyeart, J.Dairy Sci., 1987, 70, 252
22. H.L. Monaco, G. Zanotti, P. Spadon, M. Bolognesi, L. Sawyer and E.E. Eliopoulos, J.Mol.Biol., 1987, 197, 695
23. R. Aschaffenburg and J. Drewry, Biochem.J., 1957, 65, 273
24. A. Moneret-Vautrin, G. Humbert, C. Alais and J.P. Grillet, Le Lait, 1982, 62, 396
25. T.N. Koritz, S. Suzuki and R.R.A. Coomb, Int.Arch.Allergy Appl.Immun., 1987 82, 72
26. A.S. McAlpine and Sawyer,L., Biochem.Soc.Trans., 1990, 18, 879
27. H.A. McKenzie, G.B. Ralston and D.C. Shaw, Biochemistry, 1972, 11, 4539
28. C. Tanford, L.G. Bunville and Y. Nozaki, J.Am.Chem.Soc., 1959, 81, 4032
29. H.A. McKenzie, Adv.Prot.Chem., 1967, 22 147
30. P. Dunnill and D.W. Green, J.Mol.Biol., 1965, 15, 147
31. S.M. Gotham, P.J. Fryer and A.M. Pritchard,Int.J.Food Sci.Tech., 1992, 27, 313
32. H. Singh and P.F. Fox, J.Dairy Res., 1986, 53,237

33. Y.V. Griko and P.L. Privalov, Biochemistry, 1992, 31, 8810
34. X.L. Qi, S. Brownlow, C. Holt and P. Sellers, 1993, in preparation
35. H.M. Wilks, K.W. Hart, R. Feeney, C.R. Dunn, H. Muirhead, W.N. Chia, D.A. Barstow, T. Atkinson, A.R. Clarke and J.J. Holbrook, Science, 1988, 242, 1541
36. X.J. Zhang, W.A. Baase and B.W. Matthews, Biochemistry, 1991, 30, 2012
37. A.R. Clarke, D.B. Wigley, W.N. Chia, D.A. Barstow, T. Atkinson and J.J. Holbrook, Nature, 1986, 324, 99
38. D.L.D. Caspar, J. Clarage, D.M. Salunke and M. Clarage, Nature, 1988, 352, 659
39. R. Lüthy, J.U. Bowie and D. Eisenberg, Nature, 1992, 356, 83
40. C.A. Batt, L.D. Rabson, D.W.S. Wong and J.E. Kinsella, Agr.Biol.Chem., 1990, 54, 949
41. A.C. Jamieson, M.A. Vandeyar, Y.C. Kang, J.E. Kinsella and C.A. Batt, Gene, 1987, 61, 85
42. M. Silva, D.W.S. Wong and C.A. Batt, Nucl.Acid.Res., 1990, 18, 3051
43. M. Totsuka, Y. Katakura, M. Shimizu, I. Kumagai, K. Miura and S. Kaminogawa, Agric.Biol.Chem., 1990, 54, 3111
44. G.J. Paterson, PhD Thesis, University of Edinburgh, 1991
45. L. Sawyer, 'Protein Engineering: Proceedings of the AFRC Conference on Protein Engineering in the Agricultural and Food Industry', ed. P.Goodenough, CPL Press, Newbury, 1992, p116
46. S. Watkins and C.A. Batt, unpublished observations, 1993
47. Y. Cho, W. Gu., S. Watkins, S.P. Lee, J.W. Brady and C.A. Batt, Protein Engineering, 1993, in press.
48. S.P. Lee, Y. Cho and C.A. Batt, J.Agr.Food Chem., 1993, 41, 1343
49. S.P. Lee and C.A. Batt, Food Texture, 1993, 24, 73
50. S.P. Lee, D.S. Kim, S. Watkins and C.A. Batt, Biosci.Biotechnol. Biochem., 1993, in press

Functional Properties of Chhana Whey Products

Alistair S. Grandison and Anita R. Jindal

DEPARTMENT OF FOOD SCIENCE AND TECHNOLOGY, THE UNIVERSITY OF READING, WHITEKNIGHTS, READING RG6 2AP, UK

1 INTRODUCTION

Chhana is a traditional Indian product used in the confectionery industry. It is produced from cow's milk by a combination of heating to about 80^0C with acidification to approximately pH 5.4 using added food-grade acids or whey. The by-product - chhana whey - contains about 60 g kg^{-1} solids, yet is generally wasted, which leads to pollution problems. It is calculated that approximately 100,000 tonnes of whey powder could be produced annually from chhana whey in India, and the popularity of chhana-based products is spreading to other countries.

In a previous study (Jindal and Grandison, 1992) the possibility of producing chhana whey powders with protein content ranging from 2-58 %, using membrane processes (reverse osmosis, ultrafiltration and diafiltration), followed by spray drying or freeze drying, has been demonstrated. The chemical composition of the protein fractions was shown to be somewhat different to commercial cheese whey powders (Jindal and Grandison, 1993). In particular, the whey proteins of chhana whey were more highly denatured, and the relative proportion of caseins was much greater. The aim of this study was to analyse the physical properties of chhana whey powders in relation to commercial cheese whey powders, and hence to assess their potential as functional ingredients in the food industry.

2 MATERIALS

Chhana whey, derived from Channel Island breeds, was obtained from Bombay Halwa Limited (Southall, Middlesex). Commercial cheese whey powders containing 12.5, 35 and 55% protein were obtained from Carbery Milk Products Limited, Ballineen, County Cork, Ireland).

3 METHODS

Preparation of powders

Chhana whey concentrates were prepared by reverse osmosis (RO), ultrafiltration (UF) or UF followed by diafiltration, using tubular polysulphone membranes (ZF99

Table 1 Composition of chhana whey powders

Powder	Solids	Protein N	Protein*	Fat	Ash	Lactose
UF SD	944	42.0	268	172	45.0	449
DF SD	944	64.1	409	393	33.6	105
UF SD	948	55.0	351	48	57.7	454
UF SD	980	34.6	221	459	29.5	270
UF FD	965	34.1	217	453	29.0	266
DF SD	945	63.4	404	140	45.3	320
DF FD	937	62.8	401	137	44.9	ND
DF SD	953	90.7	579	113	38.2	155
DF FD	961	91.4	584	114	38.5	ND
RO SDW	956	3.3	21	60	65.2	741
RO SDS	946	3.1	20	13	65.6	828
RO FDS	922	3.0	19	13	64.0	807

All values expressed as g kg^{-1}; ND - Not Determined; * - Protein Nx 6.38
UF - Ultrafiltration; DF - Diafiltration; RO - Reverse Osmosis
SD - Spray Dried; FD - Freeze Dried
SDW - Spray Dried Whole; SDS - Spray Dried Separated; FDS - Freeze Dried Separated

RO membranes, or ES 625 UF membranes, Paterson Candy International, Whitchurch, Hants.). Powders were prepared by spray drying or freeze drying, and the gross composition of products is shown in the table 1. Fat removal by centrifugation was carried out on the wheys prior to membrane processing in most cases. Due to the variability of the efficiency of fat separation, and the high degree of concentration of the wheys, the fat levels were impossible to control accurately.

Physical Properties

Protein solubility (West et al.,1986), emulsifying activity (EA) and emulsion stability (ES) (James and Patel, 1988), foaming properties (Patel and Fry, 1987), solution viscosity (Brookfield viscometer) and heat gelation (Patel and Stripp, 1988) of chhana products were measured over the pH range 2.5-9.0, and results were compared to data for commercial cheese whey powders of similar protein contents.

4 RESULTS

Protein Solubility

Protein solubilities of chhana whey powders varied from 57-100% depending on pH. Results for some chhana and commercial cheese whey powders are shown in Figure 1. Protein solubility was lower in the isoelectric region (pH 4-5) for all powders, but was quite high outside this range. For the low protein powders produced by RO, the protein solubilities were high over the whole pH range. In general, the results for chhana and cheese whey products were very similar.

Emulsifying properties

EA (a measure of the ability to form emulsions) and ES (the stability of resulting emulsions) were measured using a standard oil-in-water emulsion test using a final protein concentration of 0.1% (w/w). Results for chhana and cheese whey products were very similar indeed over the pH range. An example is shown in Figure 2.

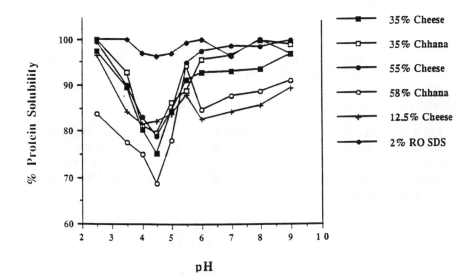

Figure 1 Comparison of protein solubilities of chhana and cheese whey powders over a range of pH.

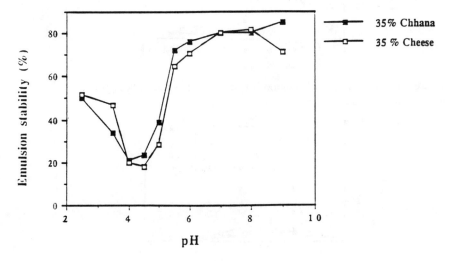

Figure 2 Emulsion stability of solutions (0.1% protein) of cheese and chhana whey protein powders.

Foaming Properties

The foaming capacity and half-life of resulting foams were measured by N_2 sparging of solutions containing 0.1% protein (w/w). Both parameters were lower for chhana powders than for equivalent cheese whey products (see Figures 3 and 4), but this was probably related to their higher fat contents. However, it was possible to produce acceptable foams using the higher protein chhana whey powders.

Viscosity

Viscosity of solutions of chhana whey powders were similar to cheese products at pH>4.0, but below 4.0 chhana products gave rise to very viscous solutions.

Gelation properties

Gel strength was measured using whey protein concentrate solutions containing 7.5% protein, heated to 80^0C for 30 min. The chhana whey protein products did not form gels at all at the higher pH values, and formed only very weak gels at the lower pH's, and were very much inferior to cheese products in this respect. Gel strengths of chhana and cheese whey powders are compared in Table 2.

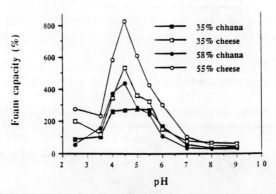

Figure 3 Foam capacities of solutions of cheese and chhana whey protein concentrates.

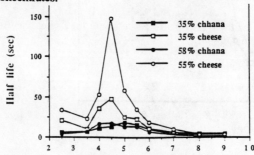

Figure 4 Stability of foams produced from cheese and chhana whey protein concentrates.

Table 2 Gel strength of chhana and cheese whey protein concentrates (7.5% solution, heated to 80^0C for 30 min)

pH	Cheese 35%	Cheese 55%	Chhana 35%	Chhana 38%	Chhana 58%
2.5	94	99	34	17	11
3.5	88	91	27	13	NM
4.0	194	201	20	NM	NM
4.5	196	207	27	NM	NM
5.0	191	199	33	NM	NM
5.5	133	142	22	NM	NM
6.0	173	181	NM	NM	NM
7.0	160	172	NM	NM	NM
8.0	291	315	NM	NM	NM
9.0	247	274	NM	NM	NM

All values expressed in g; NM - Not measurable

5 CONCLUSIONS

- Protein solubility of chhana whey powders varied from 57-100% over the pH range 2.5-9.0 and behaviour was very similar to cheese whey protein powders.

- Chhana whey powders had comparable emulsifying properties to cheese whey products.

- Foaming properties of chhana whey powders were inferior to commercial cheese whey products, but this was probably due to higher fat levels.

- Viscosity of solutions of chhana whey powders was similar to equivalent cheese products above pH 4.0, but was very high at pH 2.5-4.0.

- Chhana products formed very weak gels on heating to 80^0C at acid pH, and did not form gels at all in the range 6.0-9.0.

These studies indicate that there is considerable potential for the use of chhana whey products as functional food ingredients.

6 REFERENCES

M.J. James and P.D. Patel, British Food Manufacturing Research Association, Research report No. 631, 1988

A.R. Jindal and A.S. Grandison, J. Sci. Food Agric., 1992, 58, 511.

A.R. Jindal and A.S. Grandison, Food Chemistry, 1993, 48, 79.

P.D. Patel and J.C.Fry, In 'Developments in Food Proteins' Vol. 5, ed. B.F.J. Hudson, Elsevier Applied Science, London, 1987, p.299.

P.D. Patel and A.M. Stripp, British Food Manufacturing Research Association, Research report No. 633, 1988.

S.I. West, S.C. Lewins, R.J. Hart and C.L. Walters, British Food Manufacturing Research Association, Research report No. 560, 1986.

Thermal Aggregation of Whey Protein Concentrates under Fluid Shear Conditions

A. J. Steventon[1], A. M. Donald[2], and L. F. Gladden[2]

[1] CAVENDISH LABORATORY, UNIVERSITY OF CAMBRIDGE, MADINGLEY ROAD, CAMBRIDGE CB3 OHE, UK

[2] DEPARTMENT OF CHEMICAL ENGINEERING, UNIVERSITY OF CAMBRIDGE, PEMBROKE STREET, CAMBRIDGE CB2 3RA, UK

1 INTRODUCTION

Whey proteins are a significant source of functional protein for the food industry[1]. The thermal aggregation of these materials is particularly important, although not well understood. On the one hand, it is responsible for texture development in foods, i.e. viscosity enhancement and gelling, whilst on the other hand, it is seen both to be detrimental to other physico-chemical properties such as surfactant and solubility properties[2], and poses severe problems during processing, such as influencing protein fouling in heat exchange equipment[3].

The whey proteins are a group of proteins which include β-lactoglobulin, α-lactalbumin, bovine serum albumin, immunoglobulins, and a number of other proteins collectively termed proteose peptones[4]. β-Lactoglobulin is the major component, accounting for about 50% of the total protein, and as a result it dominates the aggregation behaviour of the whey proteins[5].

Mild heat treatment, typically around 70°C, is sufficient to induce thermal aggregation reactions between the whey proteins[6-8]. The initial effect of heat is to cause the proteins to denature, with a concomitant exposure of reactive side chains. In the denatured state the proteins behave like 'sticky' particles and are susceptible to aggregation reactions, in which they associate with one another to form high molecular weight aggregates[9]. Bonding in the aggregates consists of (i) a few strong disulphide bonds and (ii) numerous weaker physical interactions such as hydrogen, ionic and van der Waals interactions[10]. Aggregation depends on a critical balance between attractive and repulsive forces between the proteins and aggregates, which are largely influenced by the solution pH and ionic strength[12]. Under conditions where the net protein charge is minimal, such as in the narrow isoelectric pH region or at high solution ionic strength, electrostatic repulsion which would otherwise prevent aggregation is minimal. As a result aggregation is rapid, and randomly organised aggregates are formed. Conversely, when electrostatic repulsion is high, for instance outside the isoelectric pH region, aggregation is slower, resulting in the formation of highly ordered aggregates[11-15].

Under fluid shear conditions aggregation involves the following steps[16]. Initially, aggregation occurs by collisions induced by the Brownian motion of the proteins. This results in the formation of 'primary aggregates'. Aggregation continues by this mechanism until the aggregates become large enough for fluid motion to promote collisions. At which point, further aggregation is governed by shear-induced collisions. Aggregation however does not continue indefinitely; instead as the aggregates get bigger their growth rate decreases, and eventually they reach a limiting size. This behaviour has been attributed to either:

(i) A reduction in the effectiveness of collisions resulting in permanent aggregation, due to particle orientation and inter-particle repulsion effects[17].

(ii) Aggregate break-up resulting from rupture due to pressure gradients across aggregates, and erosion and fragmentation due to aggregate-aggregate and aggregate-solid surface collisions[18-20].

It is likely that both mechanisms play a part in limiting the aggregate size, but the break-up mechanisms are assumed to dominate[18-20]. The final aggregate size is therefore assumed to depend on a balance between shear-induced aggregate formation and shear-controlled aggregate break-up processes.

The overall thermal aggregation of proteins under shear conditions can therefore be considered to be a two step process, involving (i) protein denaturation, followed by (ii) an aggregation step involving shear-induced aggregate formation and break-up processes.

Textural characteristics of aggregated protein products formed under shear conditions are expected to be largely influenced by the particle size distribution. From a processing point of view, therefore, control of the size distribution is important. The study described here investigates the effects of fluid shear and process temperature on the thermal aggregation of a commercial whey protein concentrate (WPC) under fluid shear.

2 MATERIALS AND METHODS

A commercial, ultra-filtrate prepared whey protein concentrate (UF-WPC), Carbelac 35 (supplied by Carbery Milk Products Ltd, County Cork, Ireland) was used. The composition of the WPC is shown in table 1. Solutions of WPC were made up in distilled water to give a concentration of 7 %w/w whey protein in solution (natural pH = 6.3).

Table 1 WPC composition

Component	Composition /%w/w
Whey protein	35%
Lactose	50%
Fat	7
Minerals	4
Moisture	4

Shear treatment of WPC suspensions was carried out in a stainless steel couette apparatus. The apparatus consisted of two concentric cylinders, in which the inner cylinder (o.d. 106 mm) was rotated inside the stationary outer cylinder (i.d. 110 mm) giving an annular gap of 2 mm. Shear rates ($\dot{\gamma}$) could be varied between 0 and 1480 s^{-1}; variation of shear rate across the gap was calculated to be less than 4%. For the shear rates studied, Taylor analysis[21] showed that on the whole the flow was laminar, but instabilities in the form of Taylor vortices may be present. Simultaneous thermal treatment was achieved by placing the couette in a constant temperature water bath. Processing temperatures could be varied between 75 and 90°C.

WPC suspensions were processed by injecting approximately 100 ml of suspension into the annular gap and subjecting them to the desired processing temperature and shear. Aggregation was monitored by removing samples from the couette, quenching them in distilled water at room temperature to prevent further aggregation, and then measuring the particle size distributions (PSD) in a Malvern Mastersizer (laser diffraction particle sizer). The size distributions were characterised in terms of the mean particle size, $\bar{d}$, and the spread of the distribution (a measure of the polydispersity), which was expressed in terms of the coefficient of variation, C_v (C_v = standard deviation / mean size). Aggregate morphology was measured using scanning electron microscopy (SEM). Samples were prepared using a method described by Fisher and Glatz[22]. Aggregates were dried onto a glass cover slip, which was then sputter coated with gold in a Polaron E5000 Sputter Coater. Aggregates were then viewed at 30 keV in an Electroscan Environmental Scanning Electron Microscope (ESEM)[23], but used in the conventional SEM mode.

3 RESULTS AND DISCUSSION

Aggregate Microstructure

Micrographs of typical WPC aggregates viewed in the ESEM are shown in figure 1a. These have been formed by processing a 7%w/w whey protein suspension at 80°C and $\dot{\gamma}$ = 540 s^{-1} for 5 minutes. The aggregates have irregularly shaped branched structures, which are consistent with the ordered aggregation mechanism expected at this pH[11-15]. Close examination (figure 1b) of the aggregates show that they are densely packed, and formed from clusters of smaller aggregated particles (primary particles) approximately 0.2 to 0.5 μm in diameter.

(a)

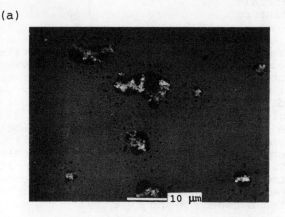

(b)

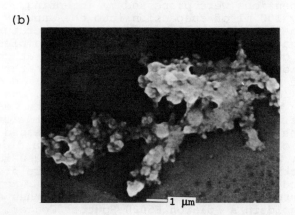

Figure 1 Typical aggregate morphology

Typical Aggregation Behaviour

Typical PSD of untreated 7%w/w whey protein suspensions, and suspensions which have been processed at 80°C and $\dot{\gamma}$ = 290 s^{-1} are shown in figure 2. The data are represented as a plot of the cumulative fraction of particles under size (CUS) versus the equivalent spherical particle diameter, d.

PSD of Untreated WPC Suspensions. In the untreated WPC suspension, some micron and larger sized particles were detected ($\bar{d}$ = 9.10 ± 2.3 µm). The existence of aggregated protein is not unexpected in a commercial preparation such as that used here. Most commercial UF-WPC powders exhibit between 30 and 50% denaturation[24], and as a result the corresponding solubility of the WPC is limited to between 77 and 80%[25], the insoluble material consisting of both lactose and aggregated protein. However, microscopical examination of the untreated solutions revealed that the small amount of micron sized particles present were crystalline, based on their sharp edges, pointing to the presence of crystals of insoluble lactose. On heating, these particles were observed to disappear and consequently the mean particle size decreased. These observations suggest that any aggregated protein present as a result of WPC preparation exists mostly as sub-micron particles, which are beyond the resolution of the Malvern Mastersizer. Therefore the particles found in the processed suspensions described below can be assumed to be formed as a result of the processing, and not artefacts resulting from the WPC preparation.

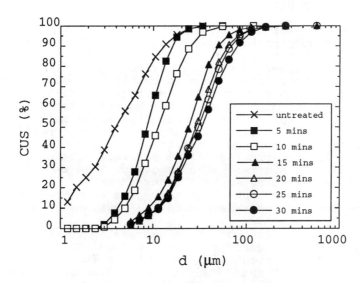

<u>Figure 2</u> Particle size distribution for aggregated WPC suspensions processed at 80°C and $\dot{\gamma}$ = 290 s^{-1}

PSD of Processed WPC Suspensions. Aggregation as a result of processing was followed by monitoring the development of the PSD in the processed suspensions (see figures 2 and 3). After 5 minutes of processing, a PSD was produced characterised by $\bar{d}$ = 9.7 µm and C_v = 0.53. Initially, the rate of aggregate growth was slow. After 10 minutes of processing, $\bar{d}$ was found to have increased only slightly up to 13.8 µm. However, for processing times between 10 and 15 minutes, there was a dramatic increase in the mean aggregate size; after 15 minutes of processing

$\bar{d}$ = 27.9 µm. For 15 < t ≤ 30 minutes aggregate size continued to increase, and after 30 minutes of processing $\bar{d}$ had increased to 43.0 µm. During this processing period however there was evidence suggesting that the rate of aggregate growth was slowing down, possibly indicating that aggregate break-up was becoming important. Concomitant with the increase in mean aggregate size during processing was a broadening of the size distribution. After 30 minutes of processing, C_v had increased to 0.87.

The time course of aggregation described above is consistent with the overall aggregation process described in the introduction. It is reasonable to assume that initially (t ≤ 10 minutes), when aggregate growth is slow, protein denaturation is the rate controlling process, whilst at longer processing times (t > 10 minutes) when aggregate growth is faster, aggregate interactions (aggregate formation and break-up) are rate controlling.

Effect of Shear Rate

The effect of shear rate on aggregation is shown in figure 3, for 7%w/w whey protein suspensions processed at 80°C. Comparison of the particles formed after 5 minutes shows that significantly larger particles were formed at 1480 s^{-1}, than at the lower shear rates. This is consistent with the higher particle collision frequencies, and thus faster aggregation rates expected at high shear rates. However, for 5 < t ≤ 30 minutes, aggregates prepared under high shear rate conditions tended to smaller sizes; the largest particles were formed under quiescent conditions, $\dot{\gamma}$ = 0 s^{-1} (see inset graph). A slight narrowing of the particle size distribution with increasing shear was also observed. This behaviour is typical of aggregating systems under shear, and is attributed to shear-controlled aggregate breakage; where aggregate breakage increases with the level of shear[22].

The susceptibility of these aggregates to shear-induced break-up was investigated further. Aggregates of mean size 43.0 ± 1.4 µm were prepared by processing 7%w/w whey protein suspensions at 80°C and $\dot{\gamma}$ = 290 s^{-1} for 30 minutes, the couette was then quenched in iced water, and the suspension subjected to a higher shear rate (1480 s^{-1}) for a further 30 minutes. The results are shown in figure 4. Aggregate size was found to decrease from 42.8 µm to 27.3 µm after shearing the suspensions for 30 minutes at the higher shear rate, suggesting that the aggregates were susceptible to shear-induced breakage. The decrease in aggregate size after switching from low to high shear conditions was accompanied initially by a broadening of the PSD, but as the time of exposure to shear increased the size distribution became narrower, as evidenced by the decrease in C_v. However, comparing the size of aggregates formed after this shear history with those obtained by processing a similar suspension at 80°C and $\dot{\gamma}$ = 1480 s^{-1} for 30 minutes (see figure 3), suggests that aggregates produced under low shear

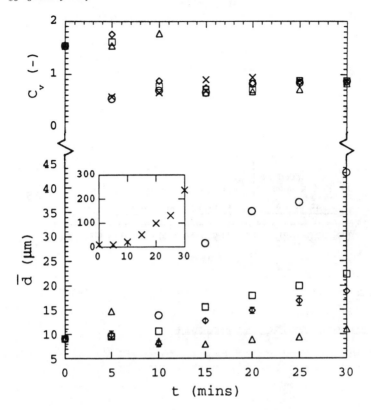

Figure 3 Effect of shear rate on aggregate size and polydispersity for WPC suspensions processed at 80°C. The inset graph shows aggregation data obtained in the absence of shear (note the difference in scale).
(× $0 \, s^{-1}$ ○ $290 \, s^{-1}$ □ $540 \, s^{-1}$ ◇ $800 \, s^{-1}$ △ $1480 \, s^{-1}$)

conditions undergo some kind of consolidation, which make them more resistant to breakage, *i.e.* the cross-links stabilising the aggregates are more effective, due to either particle packing or rearrangement.

The effect of shear rate on aggregate size described above is supportive of the idea of a balance between shear controlled growth and breakage. Initially (t ≤ 5 minutes), when the aggregates are small, break-up processes are negligible and so aggregate growth increases with shear. However, as the aggregates get bigger, aggregate break-up mechanisms become significant, and increase with fluid shear. As a result, aggregates formed under high shear conditions tend to smaller sizes than those formed in a low shear environment.

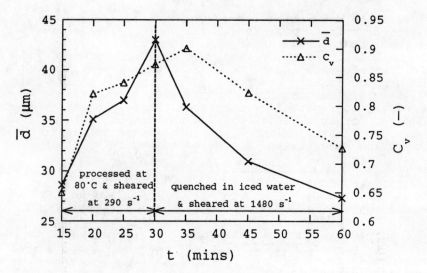

Figure 4 Effect of shear rate on aggregate stability

Effect of processing temperature

The effect of processing temperature (θ) on aggregation is shown in figure 5 for 7%w/w whey protein suspensions

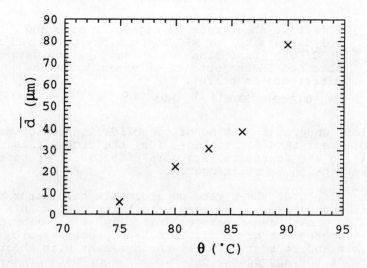

Figure 5 Effect of process temperature on aggregate size for WPC suspensions processed at $\dot{\gamma} = 540$ s^{-1} for 30 minutes

processed at $\dot{\gamma}$ = 540 s^{-1} for 30 minutes. For suspensions processed at 75°C only a small change in particle size was observed. Since the denaturation temperature for β-lactoglobulin is about 65°C, the kinetics for protein unfolding are expected to be slow at this temperature, i.e. denaturation is rate controlling over the whole time course. It is clearly seen that as the processing temperature increases, aggregate size increases. This is consistent with increased molecular unfolding and the concomitant increase in exposed cross-linking sites, which leads to faster aggregation kinetics and increased aggregate stability.

4 CONCLUSIONS

Thermal aggregation of whey protein concentrates in laminar couette flow is described. The time course of aggregation was consistent with a two step process; (i) an initial slow step where protein denaturation was assumed to be rate limiting, followed by (ii) a faster aggregate growth step, where aggregate interactions (shear-controlled aggregate growth and break-up) were assumed to be rate limiting. Aggregate growth decreased with increasing shear. The largest particles were formed under conditions of lowest fluid shear, the limiting case being in the absence of shear. The decrease in particle size with shear was attributed to aggregate break-up mechanisms. Aggregation increased with process temperature. This was attributed largely to the increased kinetics of protein unfolding, giving rise to faster aggregation kinetics and increased aggregate stability. Clearly, this work demonstrates that aggregation processes can be controlled by careful manipulation of the usual processing variables such as temperature and fluid shear, and hence desired aggregation products can be formed.

ACKNOWLEDGEMENTS

The authors are grateful to the Broodbank Fellowship Fund, Cambridge University for financial support. The authors would also like to acknowledge the DTI, Unilever, Schlumberger and ICI for funding the purchase of the ESEM, and Dr. P. Meredith and Mr. A. Eddy for help in obtaining the electron micrographs.

REFERENCES

1. R. C. Bottomley, M. T. A. Evans and C. J. Parkinson, 'Food Gels', ed. P. Harris, Elsevier Applied Science, London, 1990, Chapter 11, p 435
2. C. V. Morr, 'Developments in Dairy Chemistry. 1. Proteins', ed. P. F. Fox, Applied Science Publishers Ltd., London, 1982, Vol 1, Chapter, p 375
3. H. G. Kessler and H. J. Beyer, Int. J. Biol. Macromol., 1991, 13, 165

4. D. M. Mulvihill and M. Donovan, *Irish J. Food Sci. Tech.*, d1987, 11, 43
5. J. N. de Wit, *Neth. Milk Dairy J.*, 1981, 35, 47
6. D. G. Dalgleish, *J. Agric. Food Chem.*, 1990, 38, (11), 1995
7. Y. L. Xiong, *J. Agric. Food Chem.*, 1992, 40, (3), 380
8. N. Parris, S. G. Anema, H. Singh and L. K. Creamer, *J. Agric. Food Chem.*, 1993, 41, (3), 460
9. A. R. Hill, *Milchwissenschaft*, 1988, 43, (9), 565
10. J. E. Kinsella, 'Food Proteins', ed. P. F. Fox and J. J. Condon, Applied Science Publishers, 1982, Chapter 4, p51
11. R. Jaenicke, *J. Polymer Sci., Part C*, 1967, 16, 2143
12. E. Barbu and M. Joly, *Faraday Discuss. Chem. Soc.*, 1953, 13, 77
13. P. Kratochvíl, P. Munk and P Bartl, *Coll. Czech. Chem. Commun.*, 1961, 26, 945
14. R. H. Schmidt, 'Protein functionality in foods', ed. J. P. Cherry, ACS Symp. Ser. 147, Washington D.C., 1981, p 131
15. A. J. Steventon, PhD Dissertation, Cambridge, 1992
16. C. D. Nelson and C. E. Glatz, *Biotech. Bioeng.*, 1985, 27, 1434
17. T. G. M. van de Ven and S. G. Mason, *Colloid and Polymer Sci.*, 1977, 255, 468
18. I. Reich and R. D. Vold, *J. Phys. Chem.*, 1959, 63, 1497
19. D. S. Parker, W. J. Kaufman and D. Jenkins, *J. Sanit. Eng. Div. Proc. Am. Soc. Civ. Eng.*, 1972, 98, (SA1), 79
20. D. J. Bell and P. Dunnill, *Biotech. Bioeng.*, 1982, 24, 1271
21. G. I. Taylor, *Phil. Trans. R. Soc. London*, 1923, 223A, 289
22. R. R. Fisher and C. E. Glatz, *Biotech. Bioeng.*, 1988, 32, 777
23. G. D. Danilatos, *J. Microsc.*, 1991, 162, 391
24. C. V. Morr, *J. Dairy Sci.*, 1985, 68, 2773
25. C. V. Morr, P. E. Swenson and R. L. Richter, *J. Food Sci.*, 1973, 38, 324

Debittering of α-Casein Hydrolysates by a Fungal Peptidase

Jacqueline Gallagher, Ara D. Kanekanian, and E. Peter Evans

SCHOOL OF CONSUMER STUDIES, TOURISM AND HOSPITALITY MANAGEMENT (FOOD AND CONSUMER RESEARCH), UNIVERSITY OF WALES, CARDIFF, 66 PARK PLACE, CARDIFF CF1 3AS, UK

1 SUMMARY

Two protease preparations, papain and neutral *Bacillus* (*Bacillus subtilis*), were used individually to produce bitter α-casein hydrolysates. Each hydrolysate was separated by gel filtration chromatography (Sephadex G-25) into five fractions, which were tasted for bitterness. Reversed phase – high performance liquid chromatography (RP-HPLC) of the most bitter fraction in each case indicated the presence of late running peaks, corresponding to hydrophobic peptides.

The debittering action of a food grade fungal peptidase (*Aspergillus oryzae*) was then studied, by observing changes in the peptide profiles of the α-casein hydrolysates. After gel filtration chromatography an additional end fraction was observed, characterising a shift in the peptide composition towards smaller molecular weight peptides and amino acids. RP-HPLC of the equivalent, previously most bitter peptide fraction shows them to contain many new early running (hydrophilic) peptides, some of which may prove to be savoury in flavour. Thus the study provides evidence for the production of possibly novel savoury (hydrophilic) peptides.

2 INTRODUCTION

There are many advantages associated with the enzymic hydrolysis of food proteins such as casein, ranging from improvements in solubility and viscosity to digestibility. However a major disadvantage encountered with casein hydrolysates is bitterness. This defect in taste renders the hydrolysate unfit for addition to foods thus curbing its usefulness and wasting protein.

Bitterness in casein hydrolysates is produced by the action of the proteolytic enzyme on the α- and β- casein fractions and is associated with hydrophobic peptides.[1] Many bitter casein hydrolysates have been studied and from them bitter peptides have been isolated and identified.[2,3,4]

Debittering of casein hydrolysates by the addition of peptidases has been achieved in studies recently published and involved the use of bacterial[5] and plant[6] peptidase extracts. Although peptidases from other sources may have increased specificity and debittering properties, their immediate use in the food industry is not possible due to strict safety regulations and so food grade enzymes already available are used in current research.

This study demonstrates the use of a commercially available food grade fungal peptidase extract, from *Aspergillus oryzae*, to debitter two hydrolysates produced by the individual actions of food grade proteases on α-casein. A modification of the peptide profiling system for flavour fractions of cheese, described by Cliffe et al (1993),[7] was made for use with casein hydrolysates.

3 MATERIALS AND METHODS

Preparation of Bitter α-Casein Hydrolysates.

Whole bovine casein was obtained by precipitation at pH 4.6 from skimmed milk. It was then fractionated using the urea method of Hipp et al(1952)[8] to give an α-casein fraction that was washed thoroughly and re-precipitated with 1M acetic acid before use. A solution (2% w/v) of the α-casein fraction was prepared by dissolving in 0.05M NaOH and adjusting to the required pH with 1M HCl. Food grade Papain with an activity of 150 Xs/g* (Rhone-Poulenc, ABM Brewing and Enzymes group, Stockport, Cheshire.) was added to the solution at pH 6.5 and an enzyme:substrate ratio (E:S w/w) of 1:20. The second hydrolysate was prepared using a food grade *Bacillus* protease (PO24) extract from *Bacillus subtilis* with an activity of 140 mU/g* (Biocatalysts Ltd, Treforest, Wales, UK.) at pH 6.0 and an E:S(w/w) ratio of 1:60. Both digests were incubated at 40°C for 4 hours and the degree of hydrolysis achieved in each was measured as the 12% TCA soluble nitrogen present, given by the 2,4,6-trinitrobenzene-sulfonic acid (TNBS) assay.[9] Hydrolysis was terminated in each case by boiling the digest in a water bath for 10 minutes and this was followed by centrifugation at 4000 rpm for 15 minutes to remove any precipitate formed.

Debittering of α-Casein Hydrolysates.

The debittering of the hydrolysates involved lowering the pH to 4.3 with 1M HCl and then dividing each hydrolysate into two measured parts. The first part was called the control bitter hydrolysate (CBH) and was placed in the incubator for as long as the second part remained there, which allowed any residual protease activity to be recorded. The *Aspergillus* peptidase (P192) with an activity of 100 mU/g+ (Biocatalysts Ltd) was then added to the second part of the hydrolysate at an E:S (w/w) of 1:30 and was incubated for 17 hours at 55°C to

produce the debittered hydrolysate (DBH). Each DBH and CBH were boiled after 17 hours to stop enzyme activity and then frozen at -20°C until required.
*(One unit (U) of protease activity was defined as that amount of enzyme that will liberate one micromole of tyrosine equivalents per minute under assay conditions [pH 7.5, 37°C, 10 minutes, on 2% casein solution (w/v)]. 1 Xs/g = 2.6 mU/g.)

Preparation of the Peptide Fraction (PF).

The 70% ethanol soluble peptide fractions (PF) of both controls and debittered hydrolysates were prepared using a variation of the method developed by Cliffe et al (1993).[7] Fermentation grade ethanol (Hayman Ltd, Witham, Essex, UK.) was added to 30 ml of hydrolysate to give a final volume of 100 ml followed by stirring for 30 minutes at room temperature and then centrifuging at 4000 rpm for 15 minutes. After rotary evaporation of the supernatant at 40°C, 1 mm Hg and redissolving in 10 ml of distilled water a 3 fold concentration of the PF was achieved.

Size-Exclusion Chromatography Using Sephadex G-25 (fine).

The Sephadex G-25 was obtained from Sigma Chemical Co. Ltd., Poole, Dorset, UK. Samples (2.5 ml of the PF) were applied under eluent to the column (1.6 x 40 cm) and eluted with distilled water at a flow rate of 16 ml/hour. Fractions were collected at 20 minute intervals and the chromatogram of eluent absorbance at 280 nm, was recorded on a chart recorder. The fractions were pooled as shown in figures 1 and 2, rotary evaporated to dryness at 40°C and re-dissolved in 1 ml of distilled water. Samples for HPLC analysis were filtered through 0.2 μm filter discs.

Tasting of Fractions.

The concentrated Sephadex fractions were tasted only by two people due to the small sample size and after each sample the taster's mouth was rinsed with distilled water. Tasting results are given in terms of bitterness, with bland describing samples of no detectable taste. Samples with flavour after debittering were described as savoury if no bitterness could be tasted.

Reversed Phase High Performance Liquid Chromatography (RP-HPLC) of Fractions.

The HPLC used was a Perkin Elmer series 410 HPLC pump fitted with an Apex octadecyl 5 μ reversed phase column (4.6 x 250 mm) and a UV detector at a wavelength of 220 nm. Each sample (10 μl) was injected and eluted with 0.06% trifluoroacetic acid (TFA)/HPLC grade water as a mobile phase, at a flow rate of 1.0 ml/min. The concentration of the mobile phase modifier (0.056% TFA/HPLC grade methanol) was increased linearly from 0-71% over 55 minutes and then to 91% over a further 5 minutes.

4 RESULTS AND DISCUSSION

The TNBS assay results (Table 1) show that after 17 hours the increase in soluble nitrogen (N) concentration was 5 mM for the control bitter hydrolysate produced by papain (CBHp) and only 2 mM for the control bitter hydrolysate produced by the *bacillus* protease (CBHb). This is a small increase in soluble N when compared to the ~50 mM increase for the debittered hydrolysates (DBHp and DBHb). Thus the control hydrolysates represent very accurately the digests as they were after just 4 hours of incubation. Boiling of the hydrolysates was used to deactivate the proteases in this case, in favour of specific enzyme inhibitors because the hydrolysates were to be tasted at a later stage.

The Sephadex G-25 gel filtration chromatograms when compared show that the peptide fraction of the control hydrolysate produced by the *bacillus* protease (CPFb) after 4 hours (figure 2a) has higher peak levels than that of the control hydrolysate produced by papain (CPFp) in figure 1a. This results from a greater amount of soluble N being loaded onto the column initially, as the higher TNBS value (12% TCA soluble N concentration) in the CBHb corresponds to an equally increased amount of soluble N in the CPFb (70% EtOH soluble N concentration).[10] However both hydrolysates have five fractions occurring at almost equivalent molecular weight ranges and the tasting results presented in Table 2 show the most bitter fraction for each to be the same fraction number 3.

The RP-HPLC chromatogram figure 3A shows that fraction 3(A) of the CPFp (figure 1a) contains only late running peaks. These peaks represent late running hydrophobic peptides which are characteristic of bitter peptides when separated on RP-HPLC.[11]

<u>Table 1</u> TNBS assay results, showing the 12% TCA soluble nitrogen present in the α-casein hydrolysates produced by papain and *bacillus* protease action over time and the final debittered hydrolysates.

HYDROLYSIS TIME (HOURS)	α-CASEIN HYDROLYSATE BY PAPAIN	α-CASEIN HYDROLYSATE BY *BACILLUS* PROTEASE
	mM Glycine Equivalent Units	
0	0	0
1	10.4	13.2
2	12.4	20.9
3	14.0	24.0
4	15.8	27.9
Enzyme action terminated after 4 hours.		

DEBITTERING TIME	CONTROL (CBHp)	DEBITTERED (DBHp)	CONTROL (CBHb)	DEBITTERED (DBHb)
17 hours	20.6	67.9	29.9	83.0

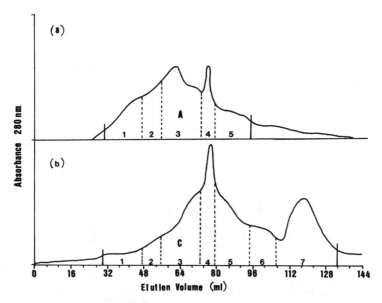

Figure 1 Sephadex G-25 gel filtration chromatograms of the Papain produced, bitter control (a) and debittered (b), α-casein hydrolysate peptide fractions.

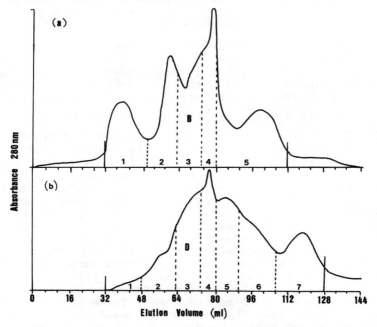

Figure 2 Sephadex G-25 gel filtration chromatograms of the *Bacillus* protease produced, bitter control (a) and debittered (b), α-casein hydrolysate peptide fractions.

Table 2 Tasting of fractions collected after Sephadex G-25 gel filtration chromatography, of the control and debittered hydrolysate peptide fractions (PF).

FRACTION NUMBER	PAPAIN, α-CASEIN HYDROLYSATE PEPTIDE FRACTION (PFp)		BACILLUS PROTEASE, α-CASEIN HYDROLYSATE PEPTIDE FRACTION (PFb)	
	CONTROL (CPFp)	DEBITTERED (DBPFp)	CONTROL (CPFb)	DEBITTERED (DBPFb)
1	bland	some flavour not bitter	bland	some flavour not bitter
2	bitter	some flavour not bitter	bitter	some flavour not bitter
3	very bitter	stronger 'savoury' flavour	extremely bitter	stronger 'savoury' flavour
4	slightly bitter	slight flavour	very bitter	slight flavour
5	bland	bland	slightly bitter	bland
6	—	bland	—	bland
7	—	bland	—	bland

Figure 3B is the RP-HPLC peptide map associated with the most bitter tasting fraction 3(B) of the CPFb gel filtration chromatogram (figure 2a) and it also shows only late running hydrophobic peptides to be present. The higher absorbance values of the peaks are consistent with there being a higher concentration of peptide material present in the CPFb.

The Sephadex G-25 gel filtration chromatogram (figure 1b), for the peptide fraction of the hydrolysate produced by papain action and then debittered by the *Aspergillus* peptidase (DBPFp), shows very clearly a shift in the peptide composition towards the later eluting material. The fraction number 5 has been confirmed by further experiment to be tyrosine and number 7 to be tryptophan. These amino acids are not eluted in order of molecular weight because the use of water as an eluent may effect ionic interactions. The presence of increased amounts of amino acids corresponds well with the large (50 mM) increase in soluble N observed in the TNBS assay results (Table 1). Upon tasting it was found that the most strongly flavoured fraction of the DBPFp was fraction 3 (Table 2), which corresponded in molecular weight range to the previously most bitter fraction. The flavour associated with this fraction could not really be described as a pleasant 'savoury' but it was however not bitter to taste. RP-HPLC analysis of the DBPFp fraction 3(C) (figure 3C) gives a peptide map showing early running peaks. These peaks represent hydrophilic peptides and their presence, combined with the

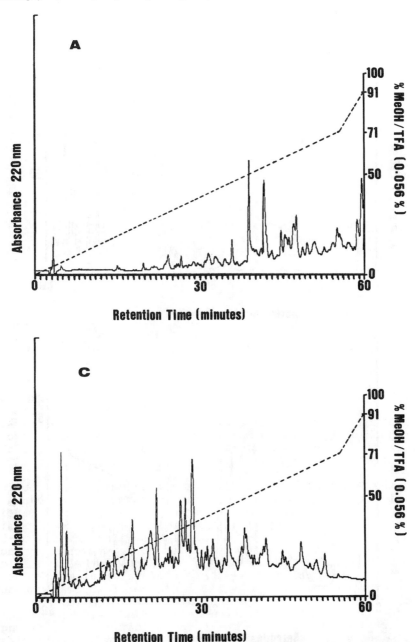

Figure 3 Reversed Phase HPLC chromatograms showing the pattern of peptides present in fraction 3 of the peptide fractions, after gel filtration. (A) Control hydrolysate produced by Papain after 4 hours. *Continued next page.*

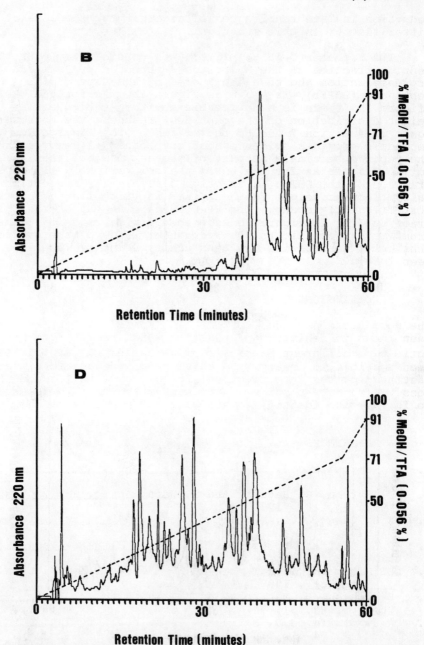

(B) Control hydrolysate produced by the *bacillus* protease after 4 hours. (C) Debittered hydrolysate produced by Papain and then *Aspergillus* peptidase after 17 hours. (D) Debittered hydrolysate produced by *bacillus* protease and then *Aspergillus* peptidase after 17 hours.

reduction in late running peaks, accounts for the loss of bitter flavour in this fraction.

The Sephadex G-25 gel filtration chromatogram, for the peptide fraction of the hydrolysate produced by *Bacillus* protease action and then debittered by the *Aspergillus* peptidase (DBPFb), is shown as figure 2b. As for the DBPFp (figure 1b) there is a noticeable shift in the molecular weight distribution of the peptides present. The appearance again of fraction 7 due to the action of the *Aspergillus* peptidase provides evidence that tryptophan is removed from the peptide terminals as part of the enzymes debittering action. Again as for the DBPFp the most strongly flavoured fraction of the DBPFb is fraction 3 (Table 2), suggesting that this molecular weight range is that in which flavour peptides most frequently occur. The RP-HPLC peptide map of the DBPFb fraction 3(D) is figure 3D and shows a wide range of peaks from the hydrophilic to the hydrophobic ends of the gradient. Thus again debittering has been achieved and new peptides have been formed.

5 CONCLUSIONS

The food grade peptidase extract of *Aspergillus oryzae* has been shown to debitter hydrolysates which result from the actions of different proteases on α-casein. By RP-HPLC it has been possible to observe the shift from hydrophobic bitter tasting peptides in a fraction, to more hydrophilic and possibly savoury peptides. The next stage of this research is to isolate and identify some of the new peptides formed.

6 REFERENCES

1. Y. Guigoz and J. Solms, Chemical Senses and Flavor, 1976, **2**, 71.
2. T. Matoba, R. Hayashi and T. Hata, Agric. Biol. Chem., 1970, **34**, (8), 1235.
3. K.M. Clegg, C.L. Lim and W. Manson, J. Dairy Res., 1974, **41**, 283.
4. R.D. Hill and H. Van Leeuwen, Aust. J. Dairy Technol., 1974, **29**, 32.
5. E. Minagawa, S. Kaminogawa, F. Tsukasaki and K. Yamauchi, J. Food Sci., 1989, **54**, (5), 1225.
6. H. Umetsu, H. Matsuoka and E. Ichishima, J. Agric. Food Chem., 1983, **31**, 50.
7. A.J. Cliffe, J.D. Marks and F. Mulholland, Int. Dairy J., 1993, **3**, (4-6), 379.
8. N.J. Hipp, M.L. Groves, J.H. Custer and T.L. McMeekin, J. Dairy Sci., 1952, **35**, 272.
9. A. Polychroniadou, J. Dairy Res., 1988, **55**, 585.
10. C.N. Kuchroo and P.F. Fox, Milchwissenschaft, 1982, **37**, (6), 331.
11. A.J. Cliffe and B.A. Law, Food Chem., 1990, **36**, 73.

The Effect of Thermisation on the Thermal Denaturation of γ-Glutamyltranspeptidase in Milk and Milk Products

Shailam S. Patel and R. Andrew Wilbey

DEPARTMENT OF FOOD SCIENCE AND TECHNOLOGY, THE UNIVERSITY OF READING, WHITEKNIGHTS, READING RG6 2AP, UK

1 INTRODUCTION

Thermolabile, psychrotrophic organisms such as *Pseudomonas* spp. can produce proteases and lipases which are extremely heat stable. High levels of psychrotrophs in raw milk are associated with the generation of off-flavours in milk and milk products, and with stability problems in UHT products.

Contamination may occur at any stage in the handling of milk, from milking on the farm to its eventual packaging in the dairy. Growth of psychrotrophs may be controlled by good plant and process hygiene, by storage at low temperature ($< 3°C$) and by limiting the storage time. Thermolabile psychrotrophs may also be killed by mild heat treatment (ie thermisation) of the raw milk. One definition of thermisation included in recent regulations [1] is: ..."thermised milk" means raw milk which has been heated for at least 15 seconds at a temperature between 57°C and 68°C and after such treatment shows a positive reaction to the phosphatase test, as described in Part IV of Schedule 5 [2]..... and for the purposes of any such test milk shall not be treated as showing a positive reaction if the sample of that milk taken for that test gives a reading of greater than 10 μ g of p-nitrophenol/ml of milk....

γ-Glutamyltranspeptidase (GGTP), EC 2.3.2.1, has been identified as a suitable enzyme for assessing heat treatments above the minimum pasteurisation conditions [3-5] for milk and cream, with denaturation characteristics similar to those of lactoperoxidase (EC 1.11.1.7). Enzyme activity is associated with membrane material and increases with the milk fat content. The rate of deactivation is a function of the heat treatment and the water activity (a_W) of the product [6].

Thermising is known to reduce the activity of alkaline phosphatase (EC 3.1.3.1) in milk. This work was carried out to investigate the effect that a thermising pretreatment has on the deactivation of GGTP in milk and milk products.

2 MATERIALS & METHODS

Raw commercial bulked milk was obtained from Cliffords Dairies Ltd (Bracknell, UK). 48% fat cream was separated from the milk at 50°C using a Lister separator and APV Junior heat exchanger (APV Baker, Peterborough) rated at 450 l/h for milk. The warm cream was standardised to 18% and used immediately, or cooled and subsequently used in the ice cream formulation as listed in table 1.

Table 1: Formulations of the milk products, expressed as g per kg

	Milk	Single Cream	Ice Cream Mix
Raw milk	1000	-	-
Skimmed milk	-	620	-
48% cream	-	380	285
Skim milk powder[†]	-	-	73
Sucrose	-	-	130
TS-D109[‡]	-	-	4.6
Water	-	-	507.4

[†] Medium heat skim milk powder, supplied by Dairy Crest Foods, Surbiton.
[‡] Emulsifier-stabiliser blend, supplied by Grindsted Products Ltd, Bury St Edmunds.

Heat treatment of the products was carried out in a second, pilot scale, APV Junior heat exchanger rated at 80 l/h. The time-temperature characteristics for product flow through the heat exchanger were characterised by a dye-injection method [7].

In the first series of treatments, the raw products were thermised by single treatments at 55°, 60° and 65°C for a minimum of 15 s then cooled to below 10°C and held in a cold store at 2°C overnight. On the following day, untreated material plus the thermised products from each pretreatment was subjected to the main heat treatments over a range of temperatures from 70° to 80°C using a mean holding time of 20.5 s (minimum hold not less than 15 s). In the second series of trials, single and double heat treatments at 65°C were carried out in a similar manner, prior to the main heat treatment. All heat treatments were carried out in triplicate.

The GGTP activity of each sample was assayed in duplicate at 37°C by measuring the release of p-nitroanilide from γ-glutamyl-p-nitroanilide in pH 8 buffer over a 5 minute period by absorbance at 410 nm[8]. Mean activities were then converted into percentages of the original activity of that untreated product, and plotted against the holding temperature for each of the heat treatments.

3 RESULTS AND DISCUSSION

The effect of the heat treatments on the activities of the milk, cream and ice cream mix are shown in Figures 1 - 3. Single heat treatments from both the pretreatments and the first series of 70°-80° heat treatments were included in the "single treatment" curve on each plot.

GGTP activity in milk (Fig. 1) was slightly reduced by the thermisation treatment (55° - 65°C for at least 15 s), with activity reduced by up to 13%. Thereafter there was a more rapid increase in denaturation with increasing holding temperature, the transition temperature corresponding to 69°C for the heat treatment conditions used in these trials. Approximately 60% of activity was lost by pasteurisation (72°C for 15s minimum). Between 70° and 76°C the reduction in activity may be represented by a linear relationship (correlation coefficient = 0.98) with activity reduced by 11.5% of the original activity per °C.

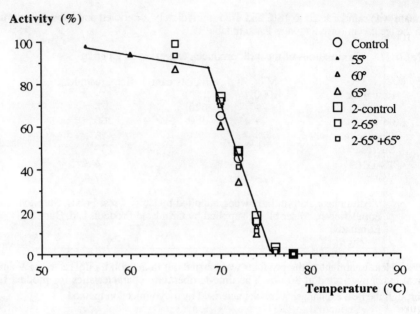

Figure 1 Reduction in GGTP activity in milk (as % of original activity) with increasing heat treatment temperature.

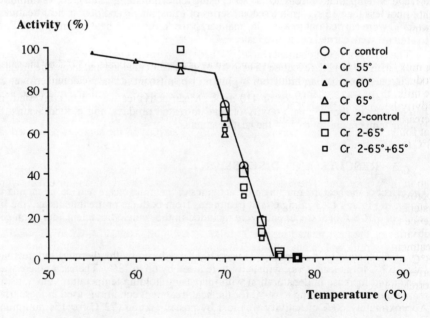

Figure 2 Reduction in GGTP activity in cream (as % of original activity) with increasing heat treatment temperature.

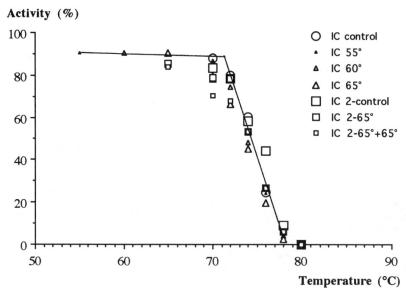

Figure 3 Reduction in GGTP activity in ice cream as % of original activity with increasing heat treatment temperature.

The pretreatments had an effect in reducing the GGTP activity at a given treatment level; the effect corresponding to the severity of the original pretreatment. In all of the trials there was a low level of activity (<1 μ mole p -nitroanilide min^{-1}) following treatment at 76°C and no activity following treatment at 78°C.

The effect of heat treatments on GGTP activity in cream (Fig. 2) was similar to that for milk, with up to 11% reduction in activity by pretreatment alone and the pretreatments producing a corresponding reduction in activity on subsequent heat treatment. As with the milk, there was a change in the rate of denaturation at 69°C, with the rate of loss of activity between 70° and 76°C corresponding to 11.5% of the original activity per °C increase in the holding temperature (correlation coefficient = 0.99) which was identical to that found for milk. There was no residual GGTP activity in the samples processed at 78°C for 15s minimum.

GGTP activity exhibited greater thermal stability in the ice cream system (Fig. 3) than in milk or cream. The single pretreatments appeared to have a similar effect, with approximately 10% loss of activity, rather than the incremental loss in the milk and cream systems. This may be attributed to the additional heat treatment during the formation of the ice cream emulsion, when the mix is prepared at 50°C prior to the heat treatments. Activities of the heat treated ice cream mixes were higher than for the corresponding treatments to milk and cream, with measurable activity remaining after heat treatment at 78°C for a minimum of 15s. Between 72° and 78°C the rate of loss of activity corresponded to 11% of the original activity per °C increase in the holding temperature (correlation coefficient = 0.98) a similar rate to that found for milk and cream.

The transition temperature in the ice cream mixes was also raised, corresponding to 71.5°C, though this was not as clear-cut as for the milk and cream. This elevation of the changeover temperature was in line with findings for heat treatment of a range of dairy products [6] where a_w was found to be the most important factor controlling the rate of deactivation of GGTP on heat treatment.

Overall, the effect of pretreatments were to make small reductions in the activity of GGTP surviving a given heat treatment, but these had no effect on the ultimate temperature at which activity was lost. Thus for a qualitative test based on the absence of activity, thermisation would appear to have no significant effect. Thermisation may however reduce the activity by approximately 10% in quantitative estimations. Below the transition temperature (69°C for milk and cream, ≈71°C for the ice cream mix) then the lower rates of deactivation must be applied. While thermisation is not permitted in the UK[9] for the heat treatment of ice cream mixes, the findings are applicable to sweetened dessert mixes.

Carter et al.[10] reported overall deactivation rates of not more than 1.1% per minute with ice cream mixes batch pasteurised at 65.6°C. Thus at temperatures at or below the transition temperature the heat treatment should be assayed by a more heat labile indicator such as alkaline phosphatase.

The authors are grateful to the Ministry of Agriculture, Fisheries & Food for their financial support of this work and for permission to publish.

REFERENCES

1. *The Milk and Dairies (Standardisation and Importation) Regulations 1992*, SI 1992 No. 3143, HMSO, London.
2. *Milk (Special Designation) Regulations 1989*, SI 1989 No. 2383, HMSO, London.
3. Andrews A.T., Anderson M. & Goodenough P.W., A study of the heat stabilities of a number of indigenous milk enzymes, *Journal of Dairy research*, 1987, **54**, 237-246.
4. Patel S.S. & Wilbey R.A., Heat exchanger performance: γ– glutamyl transpeptidase assay as a heat treatment indicator for dairy products,*Journal of the Society of Dairy Technology*, 1989, **42** (3), 79-80.8.
5. McKellar R.C., Emmons D.B. & Farber J., Gamma-glutamyl transpeptidase in milk and butter as an indicator of heat treatment, *International Dairy Journal*, 1991, **1**, 241-251.
6. Patel S.S. & Wilbey R.A., Thermal inactivation of gamma-glutamyltranspeptidase and *Enterococcus faecium* in milk-based systems, *Journal of Dairy Research*, 1994, **61** (2), in press.
7. Patel S.S. & Wilbey R.A., Heat exchanger performance: the use of food colourings for estimation of minimum residence time, *Journal of the Society of Dairy Technology*, 1990, **43** (1), pp. 25-26.
8. Baumrucker C.R., Gamma-glutamyl transpeptidase of bovine milk membranes; distribution and characterisation, *Journal of Dairy Science*, 1979, **62**, 253-258.
9. *Ice Cream (Heat Treatment etc.) Regulations 1959*, SI 1959 No. 734, HMSO, London.
10. Carter D.C., Cavanagh C.F. Higgins J.L. & Wilbey R.A., Assessment of the heat treatment of ice cream mixes by enzyme assay, *Journal of the Society of Dairy Technology*, 1990, **43** (3), 67-68.

Keeping Quality of Pasteurised and High Pasteurised Milk

B. Borde-Lekona, M. J. Lewis, and W. F. Harrigan

DEPARTMENT OF FOOD SCIENCE AND TECHNOLOGY, THE UNIVERSITY OF READING, WHITEKNIGHTS, READING RG6 2AP, UK

Introduction

There is much interest in extending the shelf-life of pasteurised milk, without inducing a cooked flavour. Factors affecting the keeping quality are: raw milk quality, heat treatment conditions, extent of post-processing contamination (PPC) and storage temperature. Of these, post-processing contamination is considered to be the most important. However if it were beneficial to do so, this could be significantly reduced. Therefore, it was felt opportune to investigate keeping quality, under conditions where procedures were used to reduce post-processing contamination to very low levels.
There is evidence about the effects of heating conditions on keeping quality. Increasing the temperature from 72 to between 80 and 95 ^{0}C or increasing the residence time from 15 to 30 or 45 s was found to reduce keeping quality (1). Also, processing conditions of 115^{0}C for 1 s was reported to activate spore germination and decrease keeping quality (2), whereas 115^{0}C for 5 s was reported to inhibit spore forming bacteria added to double cream (3). In terms of quality, it would be desirable to use heating conditions which did not induce a cooked flavour in milk.
Storage temperature will also have a marked influence on keeping quality, which improves as the storage temperature is reduced.
The aim of this work was to heat treat milk at 72^{0}C for 15 s and at 115^{0}C for 2 s, in the virtual absence of PPC and to determine its keeping quality at two different storage temperatures (2 and 10^{0}C).

Experimental Conditions

Raw milk from the same batch was processed at two sets of conditions: 72^{0}C for 15 s and 115^{0}C for 2s. This latter process was selected because it was on the threshold of cooked flavour.
An APV Junior UHT plant was used to process the milk: it has been modified to allow pasteurisation by making use of the hot water set, or higher temperatures by using pressurised steam.
Post-pasteurisation contamination was reduced by circulating hot water at about 120^{0}C for at least 10 min, through the cooling section beforehand.
All temperature instruments are checked, calibrated and certified at regular intervals.
The pasteurised milk was collected in Sterilin bottles, in a laminar air flow cabinet. Samples from each heat treatment were stored at 2^{0}C and 10^{0}C. Standard plate counts (SPC) and Aerobic spores counts (ASC) were determined

at weekly intervals, for three weeks, starting one day after production. Five separate samples were analyzed for each microbiological measurement. Once analyzed, samples were discarded.

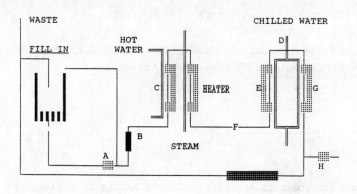

A-Feed Pump   C-Pre Heater   E-Cooler   G-Second Cooler
B-Flow meter   D-Homogenizer   F-Holding Tube   H-Final Product

FIG I PASTEURIZATION PLANT SCHEME

Results

SPC and ASC counts for raw milk are shown in Table 1

Table 1 SPC and ASC counts in raw milk, log (cfu/ml)

SPC	ASC
3.32	3.18
3.51	3.23
3.49	3.26
3.30	3.26
3.36	3.26

SPC counts and ASC counts for the heat treated milks are shown in Tables 2 and 3 for a period of 22 days. Two samples for each combination of heat treatment and storage temperature were kept for 36 days before being analyzed. The results are shown in Table 5.

Table 2 SPC counts during storage, log (cfu/ml)

		2 C				10 C		
	day 1	day 8	day 15	day 22	day 1	day 8	day 15	day 22
72 C	2,88	4,03	1,95	2,23	2,83	2,82	6,20	9,23
	2,78	2,23	2,41	2,52	2,85	2,86	5,95	11,00
	2,56	1,95	2,34	2,04	2,80	2,77	6,97	9,77
	2,66	2,34	2,28	2,23	2,92	3,08	6,98	7,28
	2,58	2,08	2,51	2,25	2,54	2,98	7,04	9,76
115 C	2,34	2,18	1,95	2,30	2,20	1,60	1,30	3,17
	2,34	1,85	1,60	1,90	2,38	1,70	1,48	2,32
	2,54	2,49	1,70	1,48	2,34	0 *	1,00	2,00
	2,36	1,78	1,00	1,48	2,11	1,30	1,60	1,48
	2,49	2,34	1,90	1,00	2,74	2,00	2,70	1,00

*no growth was observed

Table 3 ASC counts during storage, log (cfu/ml)

	2 C				10 C			
	day 1	day 8	day 15	day 22	day 1	day 8	day 15	day 22
72 C	3,32	3,35	3,36	4,03	3,37	3,48	3,40	3,08
	3,32	3,49	3,59	3,16	3,28	3,23	3,24	3,00
	3,25	3,59	3,30	3,13	3,22	3,68	3,06	3,11
	3,06	3,42	3,34	3,06	3,23	4,24	2,57	2,90
	3,26	3,40	3,26	3,17	3,29	3,78	3,01	3,32
115 C	3,05	3,18	3,00	2,77	3,31	3,18	2,23	2,70
	3,05	3,16	2,95	3,38	3,11	3,03	2,04	2,95
	3,31	2,97	3,18	3,10	3,24	3,11	2,45	2,00
	3,17	3,07	2,70	2,36	3,06	3,03	2,17	2,48
	3,20	2,76	3,26	2,78	3,12	3,15	2,41	2,85

pH values for heat treated milks are shown in Table 4.

Table 4 pH of raw milk and changes during storage

RAW	6,7	6,7	6,6	6,7	6,6

PASTEUR	115/2	115/10	72/2	72/10
day 1	6,7	6,7	6,6	6,7
	6,8	6,8	6,8	6,8
day 3	6,7	6,7	6,5	6,7
	6,7	6,6	6,6	6,6
day 7	6,8	6,7	6,6	6,5
	6,8	6,7	6,7	6,7
day 9	6,8	6,8	6,7	6,7
	6,8	6,8	6,8	6,7
day 11	6,7	6,5	6,7	6,6
	6,6	6,5	6,8	6,7
day 14	6,8	6,7	6,7	6,7
	6,8	6,7	6,7	6,6
day 16	6,7	6,7	6,6	6,4 *
	6,7	6,7	6,7	6,6
day 18	6,7	6,6	6,7	6,4
	6,7	6,7	6,7	6,5
day 21	6,7	6,6	6,7	6,1
	6,7	6,7	6,7	6,2
day 23	6,7	6,7	6,7	6,1
	6,7	6,7	6,7	6,3
day 25	6,7	6,7	6,7	6,4
	6,8	6,7	6,7	6,0
day 36	6,7	6,6	6,6	6,2
	6,6	6,6	6,6	5,9

*the milk was detected sour since this day

Discussion and Conclusions

The microbiological quality of the raw milk was excellent. The mean SPC count was 2490 / ml and the ASC was 1730 /ml.
The heat treatments used had little effect on ASC and achieved about 1 decimal reduction for SPC.
The results for ASC and SPC during storage are depicted in Figs. 2 and 3.

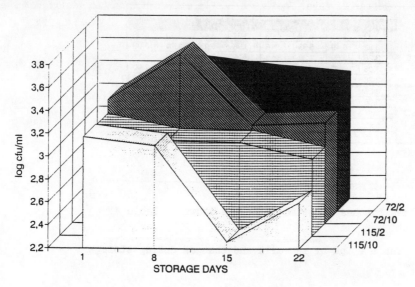

Fig 2 ASC count for milk during storage

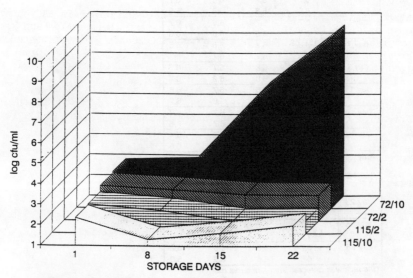

Fig 3 SPC count for milk during storage

Milk pasteurised at 72°C for 15 s and stored at 10°C became unacceptable after about 15 days. However milk stored at 2°C was still acceptable after 22 days. Two samples analyzed after 36 days also showed acceptable total counts and spore counts, with values well below 10^4/ml. This is well below 10^7/ml, which is considered to be on the threshold of acceptability.

Milk treated at 115°C for 2 s and stored both at 10°C and 2°C was still acceptable after 22 days. There was little difference in the quality of the milks (Table 5).

Table 5 SPC and ASC counts after 36 days storage

	SPC		ASC	
	2°C	10°C	2°C	10°C
72°C	1.78	XXX	3.40	XXX
	3.18	XXX	2.82	XXX
115°C	n.g.	1.30	2.91	3.06
	1.00	4.20	2.82	2.30

Samples analyzed after 36 days storage were still acceptable. However, there seemed little advantage in using the harsher heat treatment for samples stored at 2°C.

Statistical analysis of the results showed that variations in ASC counts were most affected by heat treatment conditions and storage time and less so by storage temperature. On the other hand SPC counts were influenced by all three factors.
Only gram positive bacteria were found in samples analyzed by gram staining after 22 days storage.

There is scope for improving the keeping quality by reducing post-pasteurisation contamination. Further attention should be paid to the role played by raw milk quality and storage temperature.

References

1 Kessler, H.G. and Horak, F.P., (1984), Milchwissenschaft, **39**, 451-454.
2 Guirguis, A.H., Griffiths, M.W. and Muir, D.D., (1983), Milchwissenschaft, **38**, 641-644.
3 Griffiths, M.W., Hurvois, Y., Philipps, J.D. and Muir, D.D., (1986), Milchwissenschaft, **41**, 403-405

Fouling and UHT Processing

P. Kastanas, M. J. Lewis, and A. Grandison

DEPARTMENT OF FOOD SCIENCE AND TECHNOLOGY, THE UNIVERSITY OF READING, WHITEKNIGHTS, READING RG6 2AP, UK

ABSTRACT

A miniature UHT plant has been designed and constructed, which comprises a preheater, main heater and a cooler. It has a variable flow rate and it is designed to heat products to between 100 and 150°C, in a few seconds. Flow rate and temperatures are monitored through a data logging system. A computer programme has been developed which allows real time calculations of the overall heat transfer coefficient (U). The plant can be used to monitor the extent of fouling and cleaning and the efficiency of cleaning for UHT processes.

INTRODUCTION

Fouling is the adhesion of material on surfaces of heat exchangers.

For milk there are two distinct types of deposit : Type A which is formed between 70-110°C, has high protein (60-70%), low mineral content (30-40%), is voluminous and has white colour. Type B deposit is formed at temperatures above 110°C (approximately) has high mineral content (70-80%), small amounts of protein, is brittle and brown in colour.

The main proteins involved in the formation of deposit are β-lactoglobulin and the caseins. The main minerals are calcium and phosphorus. Fat plays an insignificant role.

The factors that affect the amount of fouling can be grouped into chemical factors and operation and design factors.

Many ways to measure fouling have been developed. The most common of these follows the changes in the amount of deposit on the heated surface, the overall heat transfer coefficient (OHTC) parameters describing the flow, the thickness of the deposit or changes of the dimensions of the flow passage, pressure, temperature profile of the fouled surface.

DESIGN OBJECTIVES

Fouling is more severe at UHT temperatures compared to pasteurization temperatures. Therefore, the test rig had to be able to achieve UHT-temperatures (above 140°C) without any long- or short-term instability under non-fouling conditions (Figure 1).

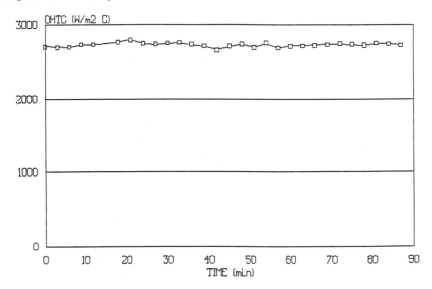

Figure 1. The Overall Heat Transfer Coefficient (OHTC) of the sterilizer when soft water is heated to a temperature 140°C proving its long- and short-term stability.

The effects of fouling of normal size exchangers often only become obvious when the plant has operated for some time (usually two or three hours). A proper investigation of the progress of the fouling process can last a few hours more. This means that a large amount of product will be necessary and therefore the cost of the fouling experiment will be high. The cost of milk for a fouling experiment with even a small commercial heat exchanger is well over £100 whereas with the miniature UHT plant it is about £10. In order to minimise the amount of product, it was decided that a maximum product flowrate of 35 kgh^{-1} was necessary.

The effects of fouling on temperature and on flowrate have to be easily measured in order to be able to measure the changes of overall heat transfer coefficient (U) which has been chosen as the characteristic parameter of the fouling process.

DESCRIPTION OF THE APPARATUS

The apparatus consists of **three tubular multipass heat exchangers** : one preheater, one heater (sterilizer) and one cooler (Figure 2).

The heating medium of the preheater is hot water, heated electrically, and the temperature is controlled with an accuracy ±1°C. The water comes from an external source and has no fouling tendencies. The dimensions of the preheater are given in Table 1. The preheater can raise the temperature of the product from ambient (15±2°C) up to 85°C when the flow rate is 15 kgh^{-1}. **The heater (sterilizer)** is heated by pressure regulated steam (0-5 bar)[1]. At the steam exit of the heat exchanger there is a thermostatic trap. It is possible to by-pass the steam-trap and use the heat exchanger at atmospheric

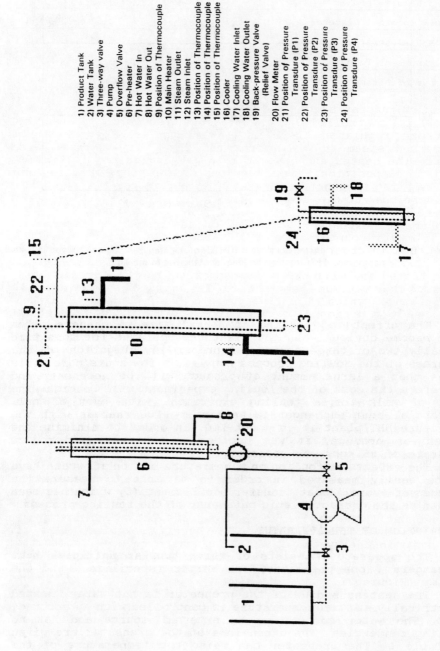

Figure 2. Block Diagram of the miniature UHT plant. The flow-path of the product is the dotted line.

1) Product Tank
2) Water Tank
3) Three-way valve
4) Pump
5) Overflow Valve
6) Pre-heater
7) Hot Water In
8) Hot Water Out
9) Position of Thermocouple
10) Main Heater
11) Steam Outlet
12) Steam Inlet
13) Position of Thermocouple
14) Position of Thermocouple
15) Position of Thermocouple
16) Cooler
17) Cooling Water Inlet
18) Cooling Water Outlet
19) Back-pressure Valve (Relief Valve)
20) Flow Meter
21) Position of Pressure Transduce (P1)
22) Position of Pressure Transduce (P2)
23) Position of Pressure Transduce (P3)
24) Position of Pressure Transduce (P4)

Table 1. Dimensions of the Heat Exchangers.

	Preheater	Heater	Cooler
Length (m)	0.445	0.985	0.48
Shell Internal Diameter (10^{-3} m)	72.3	72.5	72.3
Tube Diameter (10^{-3} m)	2.667	2.667	2.667
Thickness of Tube Wall (10^{-3} m)	0.254	0.254	0.254
Number of Passes	3	2	2

pressure, thus at a lower temperature range. The temperature of steam is controlled by a pressure regulator. Before the pressure regulator there is a venting valve which allows condensed steam to drain from the pipe. It is possible to rotate the heater and have it in a perpendicular, inclined or horizontal position. The dimensions of the heater are in Table 1. The first passage tube is extended before the heat exchanger for more than 80 diameters (20 cm) giving a fully developed flow regime in the heat exchanger. It takes approximately 90s to set up the heat exchanger depending on the operation parameters. The flow is counter-current and downwards through the first pass tube in order to reduce the formation of air pockets due to gases escaping from the milk during heating. The heater has been tested and will raise the temperature of the product up to 145°C. The maximum temperature is imposed by the maximum temperature which the instruments can handle. The heat exchanger is designed to allow tubes to be easily removed and inserted. The shell of the heater is drained before each experiment and the external surface of the tubes is tested for deposition or corrosion. The outer surface of the shell is lagged to minimise heat losses.

The **cooler** uses mains water as shell-side medium with a temperature of approximately 10(±1)°C with no fouling tendencies. The dimensions of the cooler are given in Table 1. There are stand-by tubes when more cooling surface is required. The product moves under the action of a multi-lobe positive displacement **pump** which together with a needle type **back pressure valve**, maintains positive pressure inside the heater to avoid boiling and separation of gases. Before the pump, there is a **three-way valve** which allows water/product or water/detergents to circulate through the heat exchanger from two different tanks (Figure 1). Soft water passes through the system until the desired conditions are established and then the flow switches to product. An overflow valve keeps the flowrate in the system at a set value and reduces fluctuation of the flowrate. The limiting operation pressure is 100 psi (6.9 bar) and is imposed by the plastic fittings of the flow meter.

The **pipe-run** of the system maintains the same diameter (1/8 inch, $2.667 \cdot 10^{-3}$ m) and the number of bends is kept to a minimum. The material of construction is 316 stainless steel.

INSTRUMENTATION

The temperature is monitored at the following points : (a) at the inlet of the heater (sterilizer) (b) at the exit of the heater (c) at the steam inlet of the heater (d) at the steam exit of the heater and (e) at the entrance of heating water in the preheater. The temperature at points (a),(b), (c),(d) is measured by **type T thermocouples** (copper-constantan) connected via a 0-20 mV A/D converter card with cold compensation junction to a **Data Logging System** (3D,Thinklab). All the thermocouples have a two-pin (male-female) joint which allows their easy disconnection and the manipulation of the system with the minimum strain on the thermocouples. The flow rate is monitored by a Pelton type turbine wheel **flow meter** (McMillan, 101-6P) which is connected via a 0-5 V A/D converter to the data logger[2]. The pressure of the steam is monitored by a Bourdon type pressure gauge (Bubenberg, 0-4 bar) and the pressure at the delivery of the pump by a Bourdon type pressure gauge (Bubenberg, 0-10 bar).

The thermocouples were calibrated in water-ice mixture and in boiling water-bath with the temperature measured by an accurate mercury-in-glass thermometer. The flowmeter was calibrated with water and milk at ambient temperature. An additional calibration was done with water at UHT temperature in order to check if the big difference in temperature affects the calibration line and it was found that it does not.

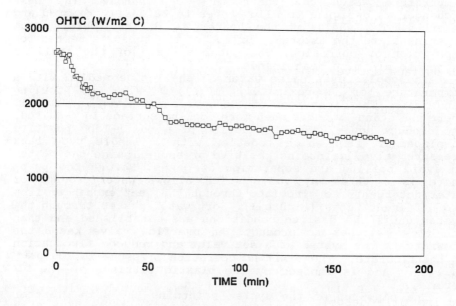

Figure 3. The Overall Heat Transfer Coefficient (OHTC) of the steriliser when reconstituted skimmed milk (pH:6.76) is heated to UHT temperature (140°C). The OHTC is calculated from Equation 1.

A specially developed data logging programme allows real time calculation of the **Overall Heat Transfer Coefficient (U)** from equation 1. Results are shown for water (Figure 1) and skimmed milk (Figure 3).

$$U = \frac{G\, C_p\, \Delta\Theta}{A\, \Delta T_{ln}} \tag{1}$$

where :
G : product mass flow rate (kgs^{-1})
C_p : Specific heat of product at product mean temperature.
$\Delta\Theta$: Change of product temperature in heater (°C)
A : Heat exchange area (m^2)
ΔT_{ln} : Logarithmic mean temperature (°C)

CLEANING

A combination of the standard cleaning procedure used in the dairy industry and the Almas-Lund protocol[3] for cleaning is used in order to achieve maximum removal of organic and inorganic deposit. At the end of each run the standard cleaning procedure is applied and before the start of each run the Almas-Lund protocol is applied.

The experiments last three hours or are stopped when the operation parameters reach limiting values. Then, the flow is switched to soft water and the steam turned off. When the temperature at the shell-side of the sterilizer has dropped below 100°C (usually after 5-10 min), the hot water to the preheater is switched on and the flow switched from water to detergent. Two kinds of detergent are used : an alkaline (NaOH based with surface active additives) and an acidic (HNO_3 based with surface active additives).The alkaline detergent is used first to remove the protein deposit. The action of alkaline detergent lasts for 30 min, its concentration is 1.5%(w/w) and its temperature is 60(±5)°C depending on the flowrate. Next, the flow is changed to acidic detergent to remove the deposited minerals. It is circulated for 30 minutes, its concentration is 1.5%(w/w) and its temperature is 60(±5)°C depending on the flowrate. Before cleaning, between the detergent stages and at the end of cleaning there are rinsing stages with soft water, lasting 10 minutes. It has been observed that the recovery of the flowrate takes place at a high rate at the beginning of the alkaline stage, reaching its maximum value at the middle of the acidic stage.

REFERENCES

1. Tissier J.P, Lalande M. (1986) "Experimental device for studying milk deposit formation on heat exchanger surface" Biotechnology Progress, **2**, 4, p.218-229.

2. Sandu C, Lund D, O'Neal B, Singh R, Almas K. (1984) "A plate heat exchanger designed to study fouling in food

processing" In: Engineering and Food, Engineering Sciences in the Food Industry, Volume I. ed. McKenna B.M. p.199-207.

3.   Almas K.A, Lund D.B. (1984) "Cleaning and Characterisation of Stainless steel Exposed to milk" Surface Technology, **23**, p.29-39.

Ultrafiltration of Sweet Cream Buttermilk

H. G. Ramachandra Rao, M. J. Lewis, and A. S. Grandison

DEPARTMENT OF FOOD SCIENCE AND TECHNOLOGY, THE UNIVERSITY OF READING, WHITEKNIGHTS, READING RG6 2AP, UK

1. INTRODUCTION

Even though buttermilk is utilised in the dairy industry as a source of SNF for replacement of skim milk, its full potential may not be realised, especially in view of its lecithin content. Membrane technologies have gained importance over the last two decades. However, the major limiting factor in the application of ultrafiltration (UF) in the dairy industry is the fall in flux with time due to concentration polarisation (CP) and fouling of the membranes[1]. Membrane fouling during UF of sweet whey has been quite widely studied[2,3]. Very limited information is available on the processing of buttermilk.

There is a distinct difference in the flux pattern between buttermilk and sweet whey. The concentration and state of some constituents such as casein, whey proteins and calcium in these products seems to be responsible for these differences. The aim of this study is to compare the fouling characteristics and CP during UF of buttermilk, skimmed milk and sweet whey and to assess the potential of UF for utilisation of buttermilk.

Objectives

- To determine the difference in flux pattern between buttermilk and sweet whey.

- To determine the rejection characteristics of the major components in buttermilk.

- To determine the effect of various components (concentration and state) on flux pattern and fouling of membranes.

2. MATERIALS AND METHODS

A Paterson Candy International tubular ultrafiltration system fitted with ES 625 membranes (polyethersulphone, area 0.8 m^2 and molecular weight cut-off 25000 dalton) was used. All the experiments were conducted in total recycle mode returning both retentate and permeate back to the feed tank.

Ultrafiltration was carried out at 50°C, inlet pressure of 0.5 MPa and outlet pressure of 0.2 MPa. The retentate flow rate was maintained at 1360 kg h^{-1}.

Rejection : The rejection of any feed component in a membrane process is defined as

$$R = \frac{Cf - CP}{Cf}$$

Cf = Concentration of a component in the feed
Cp = Concentration of a component in the permeate

Fouling Index (FI)

$$FI = \frac{\text{Water flux after rinsing following processing}}{\text{Initial flux of pure water through the membrane}}$$

Transmembrane Pressure (TMP)

This designates the pressure gradient between the upstream (retentate side) and downstream (permeate side) of the membrane. In practice, it is calculated from the average pressure at the inlet (P_1) and outlet (P_2) of the module and expressed in units of MPa.

$$TMP = \frac{P_1 + P_2}{2}$$

Pressure Drop (Δ_p) : This designates the pressure difference between inlet and outlet pressures. An increase in pressure will result from an increased flow rate of the feed stream. It is calculated from the expression

$$\Delta_p = P_1 - P_2$$

3. RESULTS

The Flux pattern of buttermilk compared to other dairy products during UF under total recycle mode is given in figure 1.

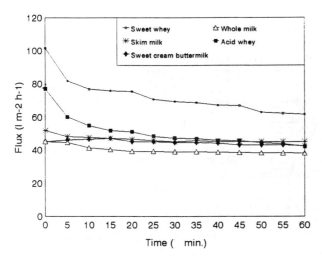

Figure 1. Flux pattern of some dairy products during UF under total recycle mode

The Flux pattern of buttermilk and sweet whey during long periods of UF under total recycle mode is given in figure 2.

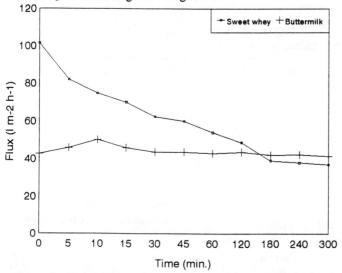

Figure 2. Flux pattern of sweet whey and buttermilk during long periods of UF under total recycle mode

FI in buttermilk and sweet whey during UF under total recycle mode is depicted in figure 3.

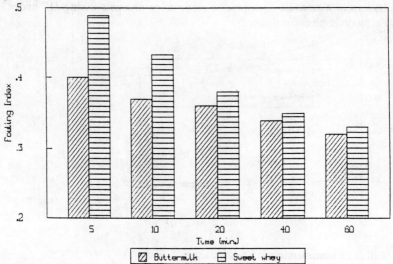

Figure 3. Fouling index in buttermilk and sweet whey during UF under total recycle mode

Effects of changing the transmembrane pressure on flux pattern of buttermilk and sweet whey are given in figures 4 and 5 respectively.

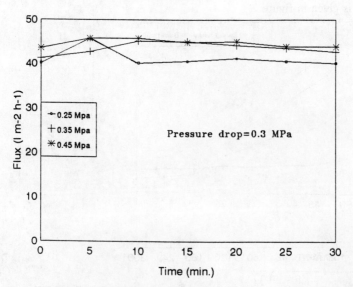

Figure 4. Effect of transmembrane pressure on flux pattern of buttermilk

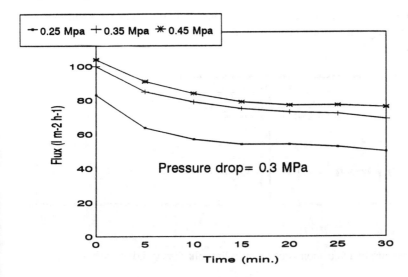

Figure 5. Effect of transmembrane pressure on flux pattern of sweet whey

Flux rates of buttermilk and skimmed milk during batch concentration are given in figure 6.

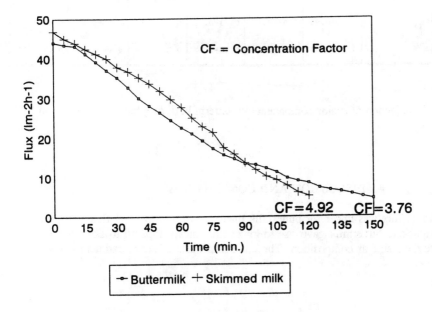

Figure 6. Batch concentration of buttermilk and skimmed milk

Rejection of major components in buttermilk during UF in total recycle and batch concentration are given in table 1 and table 2.

	Time (min.)					
	0	30	60	90	120	150
Fat	0.75-0.77	0.80-0.83	0.83-0.85	0.85-0.88	0.86-0.88	0.86-0.88
Protein	0.95-0.97	0.96-0.98	0.93-0.95	0.94-0.96	0.95-0.97	0.95-0.97
Lactose	0.29-0.32	0.1-0.14	0.10-0.12	0.08-0.10	0.09-0.11	0.08-0.11

Table 1. Rejection of major components in buttermilk during UF in total recycle mode

	Time (min.)					
	0	30	60	90	120	150
Fat	0.71-0.75	0.81-0.83	0.83-0.84	0.85-0.88	0.86-0.90	0.92-0.96
Protein	0.95-0.98	0.94-0.97	0.93-0.95	0.94-0.97	0.95-0.98	0.97-0.98
Lactose	0.48-0.50	0.23-0.27	0.15-0.22	0.09-0.14	0.06-0.10	0.01-0.02

Table 2. Rejection of major components in buttermilk during batch concentration

4. DISCUSSION AND CONCLUSIONS

The initial flux was lower for buttermilk than for sweet whey, but flux stabilised more quickly. Buttermilk gave rise to lower flux rates compared to skimmed milk despite their similar composition. The initial flux rates of sweet and acid whey were higher than buttermilk, skimmed milk and whole milk. However, the drop in flux with time was also higher in whey products. During an extended run of five hours at constant composition, the flux stabilised much faster in buttermilk than sweet whey. There was no relationship between FI and flux rates for buttermilk. However, with sweet whey, the reduction in flux rates was associated with decreased FI, suggesting that flux reduction is controlled by fouling in sweet whey. FI for buttermilk was 0.4 after 5 minutes falling to 0.32 after 60 minutes. During this time flux did not decline, suggesting that CP was controlling the process. In sweet whey, FI was 0.5 after 5 minutes with a progressive decline over the next 55 minutes. The flux also fell during this period suggesting that flux was controlled by fouling. An increase in transmembrane pressure in buttermilk form 0.25 MPa to 0.45 MPa had no positive effect on flux rates. However, in sweet whey an increase in transmembrane pressure from 0.25 MPa to 0.35 MPa increased the flux rates. The results suggest that with buttermilk the formation of a CP layer coupled with adsorption during the first few minutes of operation was mainly responsible for lower initial flux. During batch concentration, the rate of concentration of buttermilk was lower than for skimmed milk. The CF achieved in buttermilk during concentration was comparatively smaller because of lower flux rates. Rejection of lactose was unusually high (>0.3) in buttermilk at the beginning of the process, but was reduced to 0.1 after 1 h operation.

5. REFERENCES

1. J.L. Maubois. J. Soc. Dairy Tech., 1980, 33, 55.

2. R.S. Patel and H. Reuter, Milchwissenchaft, 1985, 40, 730.

3. P.S. Tong, D.M. Barbano, and M.A. Rudan, J. Dairy Sci., 1988, 71, 604.

Subject Index

Acid proteinase, milk, 2
Adjunct cheese starters, 1,10
  peptidases, 10
  proteinases, 10
Affinity chromatography
  recombinant chymosin, 78
Alkaline phosphatase, 152,156
Amino acids
  in cheese, 1,33,49
  in milk, 32
6-Aminohexanoic acid, 2
Aspartic proteinases, 73
  loop structures, 73-75,80-82
Aspergillus oryzae peptidase, 143,144, 150

Bacillus subtilis proteinase, 143,144, 147,150
Bacteriophage, 47-49
Bitter peptides, 6,49,50,58,67,143-151
Bovine serum albumin, 94,104
Brevibacterium linens, 1

Carboxypeptidase, 15,34
Casein
  bioactive peptides, 98,108
  bitter peptides, 143-151
    debittering, 143,144,146
    fractionation, 145,146,148
  cryodestabilisation, 96
  drying, 96
  emulsifying properties, 97
  ethanol precipitation, 96
  foaming properties, 97
  fractionation, 97
  genetic variants, 95
  human milk, 97
  hydrolysates, 143-146
  hydrophilic peptides, 148

Casein (continued)
  industrial production, 95,97
  micelles, 3,95,97
  milling, 96
  rennet, 96
  solubility, 95
  thermal stability, 95,98
Caseins (bovine)
  amino acid sequences, 7-8
  α_{s1}- , 2,3,4,5,6,15,26,42,84,94,143
  α_{s2}- , 3,4,5,6,84,94
  β- , 2-6,15,16,26,42,67,84,94, 97-98,143
  δ- , 6
  κ- , 3,4,6,42,84,94,96,98
  γ- , 2,5,15
  para-κ- , 3,4,96
β-Casomorphins, 98,108
Cathepsin D, 2,5,6,73
Cheese
  adjunct starters, 1,10
  amino acids in, 1,33,49
  bitter peptides, 6,49,50,58,143-151
  coagulant, 1,3,4,50,72-82,96,98
  curd, 1,96,98
  flavour, 41,43,47,49-51,58,83, 84,88
  non-starter flora, 1,2,10,14,23,24,26
  peptides, 1,15-22,25,144
    fractionation, 15-22,144
  ripening
    biochemistry of, 1,53
    flavour development, 83,88
    glycolysis in, 1,95
    lipolysis, 1
    proteolysis in, 1-10,15-22,26,50
  starter bacteria, 1,32-44,47-53,95
  varieties
    blue mould, 1

Cheese (continued)
  varieties (continued)
    Camembert, 14
    Cheddar, 2,3,10,15-26,14,47
    Dutch, 3,47
    Gouda, 2,10
    Meshanger, 2
    Mozzarella, 1,2
    surface ripened, 3,14
    Swiss, 2,47
  whey, 94,96,98-106
Chhana whey, 127-132
  concentrates, 127,128
  physical properties, 128-131
  powders, 127
    comparison with cheese whey, 129-131
    composition, 128
    emulsifying properties, 129
    foaming properties, 130
    gelation properties, 130
    solubility, 128,129
    viscosity, 130
  utilisation, 127
Chymosin, 3,49,50,72-82
  action in cheese, 15,94
  action on caseins, 3,94
  alternative coagulants, 4,73
  amino acid sequence, 75
  combined action with plasmin, 6
  DNA sequencing, 75,76
  expression in *Trichoderma reesei*, 75-79
  loop structures, 73-75,80-82
  protein engineering, 72-82
  recombinant, 4,15,79-82
    analysis of, 79
    crystallisation of, 79
    purification, 78
  specificity, 3,15,74
  stability, 74
  structure, 72-75,80-82
  X-ray crystallography, 80
Coagulants (*see also* chymosin, rennet)
  microbial, 4,73,75
  residual, in cheese, 4
Colloidal calcium phosphate, 95
Colostrum, 105,106
Coprecipitate, 107

Diafiltration, 101,102,107,127
Dipeptidase, 13-14,34,35,87,89

Electrodialysis, 97,100

Fermented dairy products, 41
Flavour peptides, 50,51,56,58,83,88, 143,144,148
Fungal peptidase, 143

Gel filtration
  bitter peptides, 143,145,147
  cheese peptides, 16,19,21,144
  recombinant chymosin, 78
Genes in lactic bacteria
  Campbell-type recombination, 39,41
  chromosomal integration, 39
  expression, 39,41,59,62-66
  food-grade selection markers, 40,42
  replacement recombination in, 40
  specifying peptidases, 37-39,66-68
  specifying proteinases, 37,59,62-66
Gluconic acid-δ-lactone, 1,10
γ-Glutamyltranspeptidase, 152-156
Glycomacropeptide, 94,96,105,106

HPLC
  bitter peptides, 143,145
  casein hydrolysates, 143,149
  cheese peptides, 15,20-22,50

Immunoglobulins, 94,104-106
  role in infant feeds, 106
  role in infections, 106
Infant feed formulas, 105,106
  action on caseins, 4
Ion-exchange chromatography
  cheese peptides, 18
  recombinant chymosin, 78
  whey proteins, 101,104

Lactalbumin, 103
α-Lactalbumin, 94,104
Lactic acid, 47,95
Lactobacillus, 7,9,12-14,47,86
Lactococcus lactis, 7,8
  aminopeptidase, 34,35,66-68,83-91
  cell lysis, 22,52
  dipeptidase, 13,35

Lactococcus lactis (continued)
- endopeptidase, 10-11,34,35
- enzyme engineering in, 32-44,56-68
- growth requirements, 32,42-44,84,90
- peptidases, 9-11,22,26,33,51,83-91
  - debittering action, 67,88
  - genetic engineering, 66-68
  - location of, 36
  - specificity of, 9-10,42,67,85-89
- peptidase specifying genes, 37-39,66
- peptide transport, 10,34,56,58,89
- proteinases, cell-wall (PrtP), 7,8,15, 33-35,39,41,56-68,91
  - catalytic domain, 61-63
  - function, 60-62
  - homology to subtilisin, 61,62
  - inactivation, 63
  - inhibitors, 63
  - loop structures, 61,63,65,66
  - membrane anchor, 61
  - P_I-type, 7,8,9,42,59,62,64
  - P_{III}-type, 7,8,9,16,42,59,60,62,64
  - P_I/P_{III}-type, 7,8
  - purification, 63
  - release from cell wall, 63,64
  - site directed mutagenesis, 62-66
  - specificity, 61,62,64,65,84,85
  - structure, 60
  - substrate binding sites, 64,65
  - thermostability, 64
- proteinase specifying genes, 37,59, 62-66
- proteolytic system, 33,57
- tripeptidase, 10,13,35,36

β-Lactoglobulin, 94,104,114-126
- acid precipitation, 123
- aggregation mechanism, 121
- chymosin cleavage, 123
- conformational structure, 116
- crystal structure, 117,118
- denaturation, 116,121,122
- functional property prediction, 124
- gelation, 121,123
- incorporation into cheese curd, 116
- *in vivo* function, 114
- lipocalin fold, 115
- loop structures, 118
- molecular structure, 115-124
- primary structure, 116

β-Lactoglobulin (continued)
- retinol binding, 114,115,120
- secondary structure, 118,119
- site-directed mutagenesis, 114-126
  - cloning and expression, 119-120
  - conformational stability, 122
  - gelation, 123
  - inclusion bodies, 119
  - objectives, 117
  - role in drug delivery, 120
  - sulphydryl groupings, 121,122
  - thermostability, 122,123
- stability to acid, 116
- sulphydryl interactions, 121
- tertiary structure, 118
- thermostability, 122,123
- yoghurt properties, 123,124

Lactoperoxidase, 94,105
- antibacterial properties, 105

Lactose, 33,47,95,96,98-100,102
- crystallisation, 100

Lactotransferrin, 94,105,106
LepI, 34,35,87
Leucocyte proteinases, 6
Lysine aminopeptidase, 6
Lysozyme, 52

Mesophilic lactobacilli, 14
Microfiltration, 97,98,101
Milk
- pasteurised
  - cooked flavour, 157
  - heat treatments, 157
  - keeping quality, 157-161
  - microbiological quality, 157-161
  - pasteurisation plant, 158
  - pH changes, 159
  - post-processing contamination, 157
  - shelf life, 157-161
  - storage temperature, 157-161
- UHT processed
  - deposit composition, 162
  - deposit measurement, 162
  - heat exchanger fouling, 162-163
  - miniture processing plant, 162-167
  - plant cleaning, 167
  - plant performance, 166,167
Milk protein
- composition, 94

Milk protein (continued)
  concentrate, 96,107
  functional properties, 94,107,108
  genetic engineering, 108
  genetic variants, 94
  heterogeneity, 94
  hydrolysates, 108,143
  modification, 107
Milk retinol-binding protein, 114
Mucor pusillus, 73
Nisin, 37,56,57
NisP, 35,37,57

Papain, 143,144,147,149
PCP, 35,36,87
Penicillium camemberti, 1
Penicillium roqueforti, 1
PepA, 11,14,35,36,87,88
PepC, 11,14,34,35,37,87,88
PepN, 10,11,14,22,34,35,38,56,58,
    66-68,87
PepO, 14,22,34,35,38,43,87
PepP, 35,36
Pepsin, 4
PepT, 10,13,14,34-36,43
Peptide transport, 10,34,56,58,89
PepX (PepXP), 10,11,22,35,36,37,42,
    43,86,87
Phenylketonuria, 106
PIP, 13,14,35,36,87
Plasmin, 2,5,16,22
  action on caseins, 5-6
  combined action with chymosin, 6
  inhibitors, 2
  specificity, 5,6,15
Prolidase, 13,14,35,36,86,89
Proteinase (*see also* PrtP, plasmin)
  acid,im milk, 2
  aspartic, 73
  genetic engineering, 74
  indiginous milk, 5,6,95
  loop structures, 73-75,80-82
  microbial, 4
  starter bacteria, 7-9,15,16,49,73-82
Proteose-peptones, 5,16,94
PrtM, 33,34,39,59,60
PrtP, 7,8,15,33-35,39,41,42,56-68,91
Psychrotrophic microorganisms, 152
  control of growth, 152

Psychrotrophic microorganisms (cont.)
  off-flavour production, 152
  stability of UHT milk, 152

Rennet, 3,4,95,98,106
Reverse osmosis, 99,127

Sodium caseinate, 97
Spherosil process, 103
Starter
  bacteria, 1,32-44,47-53,95
  bacteriophage, 47-49,52
  cultures, 47-53,58,83
  lactic acid production, 47,48,56
  proteinases, 7-9,15,49
    inactivation of, 63
    purification, 63
    site-directed mutagenesis, 7,32-44,
      52,53,56-68
    specificity of, 7-9,15,61-64,84,85
  peptidases, 2,9-14,42,49
    functions of, 42,67
    site-directed mutagenesis, 32-44,
      52,53,56-68
Streptococcus, 7,10,47

Thermisation, 152-156
  cream, 153-155
  ice-cream mix, 153,155,156
  milk, 152-155
  sweetened desert mixes, 156
Thrombin, 6
Tripeptidase (PepT), 10,13,14,34-36,
    38,43,87,89

Ultracentrifugation, 97,101
Ultrafiltration
  buttermilk, 169,171-175
  caseinate, 98
  concentration polarisation, 169,173
  fouling index, 170,172,175
  membrane flux, 169-175
  membrane fouling, 101,133,169
  permeate, 170
  retentate, 170
  skimmilk, 96,107,169,171,174,175
  whey, 101,102,104,127,169,171-175
  whole milk, 171,175

Vistec process, 102

Whey
  acid, 98
  cheese, 94,96,98-106
  composition, 99,133
  demineralisation, 100,104
  dried, 98-101
  powder, 99-101
  processing, 99
  reverse osmosis, 99
  sweet, 98
  syneresis, 124
  uses, 99
Whey protein, 94-97
  aggregation, 133
    effect of shear, 134,138-141
    mechanism, 134
    microstructure, 135
    particle growth, 137,138
    particle size, 138-141
    polydispersity, 137,139
    primary particles, 136
    stability, 138-140
    temperature effects, 140
  allergenicity, 104
  bioactive peptides, 108
  concentrate, 101-103,124,133-142
  denaturation, 103,108,133
  disaggregation, 134,138
  drying, 102,103
  fat substitute, 108
  fractionation, 104-106
  functional properties, 133
  isolate, 101,102
  pressure, effects of, 104
  products, 98
  properties of, 104,123,124,133
  solubility, 95,133
  thermal stability, 95,103,133,140,141